STUDENT SOLUTIONS MANUAL

To Accompany

ELEMENTARY STATISTICS

From Discovery to Decision

MARILYN K. PELOSI
Western New England College

THERESA M. SANDIFER
Southern Connecticut State University

WILEY

JOHN WILEY & SONS, INC.

Cover photo: Portland Japanese Garden by Tom Wilkins

To order books or for customer service call 1-800-CALL-WILEY (225-5945).

ISBN 0-471-26709-0

10 9 8 7 6 5 4 3 2 1

Table of Contents

Chapter 1 The Language of Statistics - Solutions

Section 1.2 Exercises

1.1   a) College students
b) Too much time, too much money, can't identify all the members.
c) Major stress, age, gender, etc.

1.3   a) There might be two different populations of interest. You could be interested in current and former students who were football players. You might also use high school seniors from schools that usually send students to the school, although that would not answer the retention questions.
b) I would imagine that faculty would have different opinions about the problems than students would. Also, all students are not looking for the same things from a college athletic program.
c) He or she would be interested so that they could adjust their recruitment techniques. Also, they would find out what they could do to keep the students in the program once they get them.

1.5   a) People who travel internationally.
b) Airport or Level of Satisfaction (on a scale) with a list of different airports
c) Passenger lists might have been obtained from different airlines and surveys sent out. Travel agencies might have been used to give surveys to their customers.

1.7   Other factors might be level of experience (how many years the person has been working), level of education, location in the country and many others.

Section 1.3 Exercises

1.9   a) These are statistics because there was sampling involved.
b) Age of patients, gender, other heath factors.

1.11   a) The parameter of interest might be the mean, the maximum or the minimum spent.
b) It is most likely a statistic.
c) Not really - knowing what other people pay won't help me.

1.13   a) The parameter of interest is teacher's salary - highest, lowest, average, median.
b) Highest salary, lowest salary, average salary.

Section 1.5 Exercises

1.15   a) Because the Friday workers might be better or worse than other workers.
b) Select 30 cartons from different days.

1.17   a) Because the people who actually use the center will probably like it.
b) Survey students throughout the campus whether they use the center or not.

1.19   Unless you intend to drive exactly the same way your friend does, the information is not useful. It does not look at all of the different ways that people would drive the car. You might as well just believe the manufacturer's specifications.

1.21   a) Because the school might be a better or worse nursing school than most. (This is probably a beneficial bias, however.)
b) Select nurses from various schools.

Section 1.6 Exercises

1.23   a) Quantity of ice cream in a carton.
        b) Quantitative
        c) Continuous

1.25   a) Qualitative
        b) Nominal

1.27   a) Qualitative, nominal
        b) Qualitative, nominal
        c) Qualitative, ordinal
        d) How many emails do you receive a day? How many (or what percent) of those e-mails do you carefully examine before opening attachments? This would allow you do define the categories from c) in different ways. Also, you can do more with quantitative variables.
        e) Qualitative, discrete

Section 1.10 Exercises

1.29   a) $\sum_{i=1}^{10} x_i$

        b) $\sum_{i=1}^{10} (x_i - 43)$

        c) -27

Chapter 1 Exercises
Learning It

1.31   a) All students in the state who are above the age of the DARE program.
        b) Too much time, money, can't identify everyone in the population.
        c) Age, gender, whether or not they completed the DARE program, school, whether or not the person takes recreational drugs or uses alcohol.
        d) Sample users

1.33   a) Because you will probably not see all of the different gasoline companies, because it is only relevant to your town and if you buy gas in other places that will not help.
        b) You could pick 10 different towns and then pick a station at random from each town.

1.35   a) $\sum_{i=1}^{7} x_i$

        b) $\sum_{i=1}^{7} (x_i - 6.7)$

        c) 1%

Thinking About It

1.37   Your information would be completely and 100% accurate, assuming that your surveying techniques were flawless.

1.39   No, it is not likely.

1.41   They only surveyed the people who had telephones. Since, in 1945, the telephone was not owned by everyone, they were probably only reaching the upper classes, and, as a consequence, they had a highly biased sample.

Chapter 2 Graphical Displays of Data - Solutions

Section 2.2 Exercises

2.1

a)

Mode of Transportation	Frequency	Relative Frequency	Cumulative Relative Frequency.
Car	17	68.00%	68.00%
Walk	3	12.00%	80.00%
Bicycle	1	4.00%	84.00%
Public	3	12.00%	96.00%
Other	1	4.00%	100.00%

a) Cars were most popular, bicycles and other tied for least popular.

b) The cars had the majority.

2.3

a)

Defect	Frequency	Relative Frequency
No Defect	8	40.0%
Unsealed	2	10.0%
Dented	4	20.0%
Crushed	6	30.0%

b) It appears that 60% of the boxes are unsealed, dented or crushed - which the customers won't buy. Although 40% of the packages do not have do not have any defect, that is not much consolation. There is reason to complain.

2.5

a)

Times per week	Frequency	Relative Frequency	Cumulative Relative Frequency
0	4	11.43%	11.43%
1	8	22.86%	34.29%
2	13	37.14%	71.43%
3	8	22.86%	94.29%
4	1	2.86%	97.14%
5	0	0.00%	97.14%
6	0	0.00%	97.14%
7	1	2.86%	100.00%

b) About 29%

c) About 6%

d) No they don't, it is only about 6%.

2.7

a)

Rating of Food	Frequency	Relative Frequency	Cumulative Relative Frequency
1	8	26.67%	26.67%
2	3	10.00%	36.67%
3	12	40.00%	76.67%
4	7	23.33%	100.00%
5	0	0.00%	100.00%

b) The highest frequency was a 3, the lowest was a 5.

c) I think the students are more or less indifferent towards the food, though the vast majority of the school definitely doesn't like it, absolutely no one loves it and 37% of the school actively dislikes it. So, judging by the fact that it's a college cafeteria, keep them, since you're not going to do much better. Food is rarely what attracts students to a college.

2.9   a) Since n = 40 and $\sqrt{40}$ is between 6 and 7 aim for 6 classes. The class width should be about $\frac{7.4-0.6}{6}=1.13$.

It would make sense to start at 0 and use a class width of 1.0.

Turn Around Time	Frequency	Relative Frequency	Cumulative Relative Frequency
$0.0 < x \le 1.0$	1	2.5	2.5
$1.0 < x \le 2.0$	4	10.0	12.5
$2.0 < x \le 3.0$	9	22.5	35.0
$3.0 < x \le 4.0$	9	22.5	57.5
$4.0 < x \le 5.0$	11	27.5	85.0
$5.0 < x \le 6.0$	5	12.5	97.5
$6.0 < x \le 7.0$	0	0.0	97.5
$7.0 < x \le 8.0$	1	2.5	100.0

b) 65%

c) Less than one hour is 2.5%. Between 2 and 5 technically does not include 2 and 5. If we add the classes from 2 to 3, 3 to 4 and 4 to 5 we get 22.5 + 22.5 + 27.5 which is 72.5%. This includes 2 values of 5 exactly and so it is a little high. The exact answer is 27/40 or 67.5%.

d) Only 1 person, for 2.5%.

e) How large is your home? Do you do anything else while you clean (like laundry or watch television)? Do you hire somebody to clean your home? How much time do you spend picking up clutter around your home?

2.11   a) $\sqrt{45}$ is between 6 and 7, so use 6

$\frac{15.2-8.3}{6}=1.15$ so in this case I will use 1 and start just below the first data point of 8.2

Height	Frequency	Relative Frequency	Cumulative Relative Frequency
$8.2 < x \le 9.2$	4	8.89%	8.89%
$9.2 < x \le 10.2$	6	13.33%	22.22%
$10.2 < x \le 11.2$	4	8.89%	31.11%
$11.2 < x \le 12.2$	5	11.11%	42.22%
$12.2 < x \le 13.2$	9	20.00%	62.22%
$13.2 < x \le 14.2$	10	22.22%	84.44%
$14.2 < x \le 15.2$	7	15.56%	100.00%

b)   Redo so that you can answer questions about heights by making classes begin and end on a whole number of inches.

Height	Frequency	Relative Frequency (%)	Cumulative Relative Frequency
$8.0 < x \le 9.0$	3	6.7%	6.7%
$9.0 < x \le 10.0$	5	11.1%	17.8%
$10.0 < x \le 11.0$	5	11.1%	28.9%
$11.0 < x \le 12.0$	4	8.9%	37.8%
$12.0 < x \le 13.0$	10	22.2%	60.0%
$13.0 < x \le 14.0$	10	22.2%	82.2%
$14.0 < x \le 15.0$	7	15.6%	97.8%

| 15.0 < x ≤ 16.0 | 1 | 2.2% | 100.0% |

c) About 62.2% (22.2+22.2+15.6+2.2)
d) About 17.8% (6.7+11.1)
e) About 64.4% (11.1+8.9+22.2+22.2)

Section 2.3 Exercises

2.13 a)

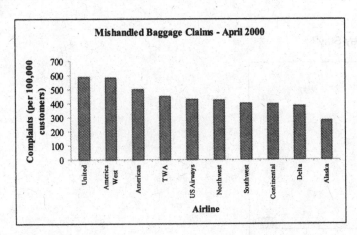

b) United and America West have the most complaints. The next seven airlines are about the same and then Alaska has the least number.

2.15

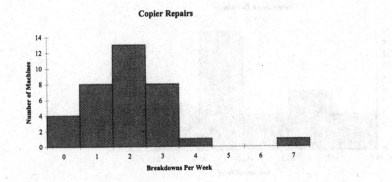

2.17 a)

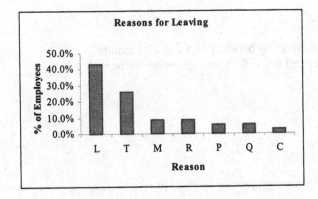

Note: A legend for the codes would also be a good idea.
b) Location is the highest frequency.
c) The first two categories make up about 65% of the total.
d) I don't see what the manager can do about location - it would be very costly for them to relocate the company and there is no reason to believe that another location would make everyone happy. The manager could address travel problems though by surveying employees to see what the problems are and creating a task force to study the problem.

2.19

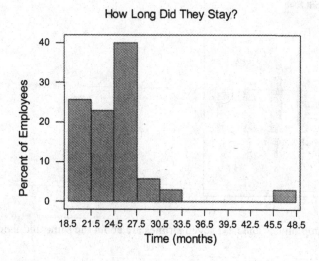

2.21

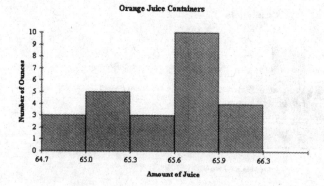

Section 2.4 Exercises

2.23   Look at the histogram from 2.19

a) A typical salesperson stayed with the company between 24.5 and 27.5 months.
b) It is not very symmetric, possibly skewed left with an unusual value on the high side.
c) The data are highly variable.

2.25 A histogram of the data:

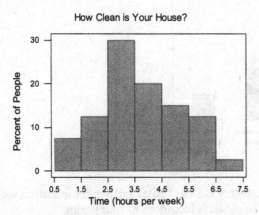

How Clean is Your House?

a) Typically the mount of time spent cleaning is between 2.5 and 3.5 hours per week.
b) The data are skewed right, meaning there is more variability on the high side of the center.
c) The times are not very variable and range from 0.5 to 7.5 hours.

Chapter 2 Exercises
Learning It

2.27 a)

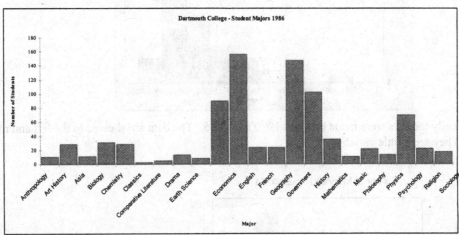

Dartmouth College - Student Majors 1986

a) About 13% majored in the hard sciences, counting mathematics.
b) and c) No matter how you do this it is going to be big. Given the absence of any other information, alphabetical works. Descending order might be good. This is a good time to collapse categories, perhaps by type of major, Sciences, Social Sciences, Arts, etc.

2.29   a)

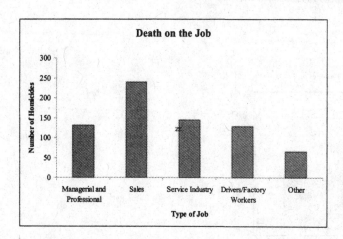

b) There is no majority, although the highest frequency category is Sales. Sales and Service industries together form a majority, which might indicate that dealing with the public is hazardous to your health.

2.31   a)

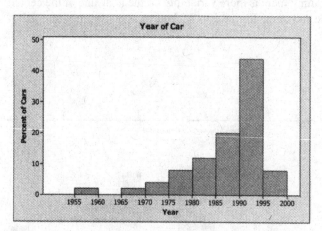

b) Typically the cars were made between 1990 and 1995. The data are skewed to the left and range from 1955 to 2000. They are a little variable.

2.33   a)

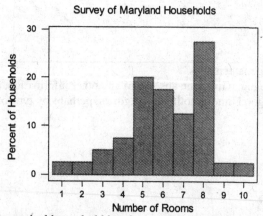

b) The typical household has 8 rooms, although there is another center at 5 and 6. The data are skewed left and highly variable.

c) According to this data, there are about 33% of the households with 8 rooms or more.

2.35   a)

Clothing $	Frequency	Relative Frequency	Cumulative %
$-50 < x \le 50$	26	65.0%	65.00%
$50 < x \le 150$	5	12.5%	77.50%
$150 < x \le 250$	4	10.0%	87.50%
$250 < x \le 350$	2	5.0%	92.50%
$350 < x \le 450$	2	5.0%	97.50%
$450 < x \le 550$	0	0.0%	97.50%
$550 < x \le 650$	0	0.0%	97.50%
$650 < x \le 750$	1	2.5%	100.00%
Total	40		

b)

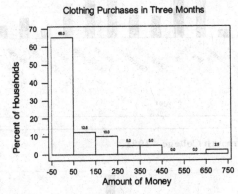

c) A majority of the households spent $50 or less per month. The data are skewed right, with a clearly unusual value between $650 and $750.

Thinking About It

2.37   a)

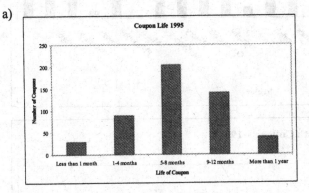 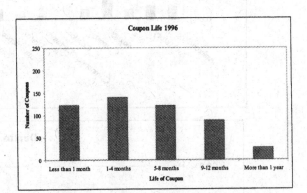

b) In 1995 the most frequent coupon life was 5-8 months, while in 1996 it was 1-4 months. In 1995, about 180 (over a third) of the coupons were for 9 months or more, while in 1996 only 115 (about a fourth) were.

2.39   a) Since the total number of students differ, you should probably be a switch to relative frequency. The totals are close, so if you use frequency you need to make sure the scales are the same.

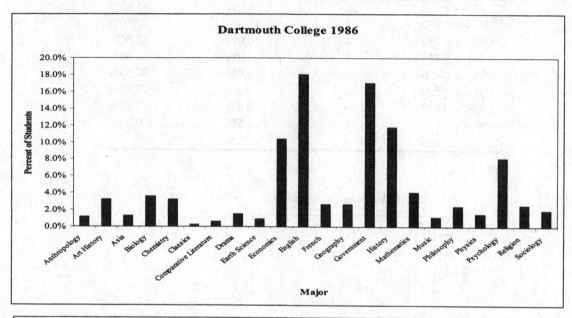

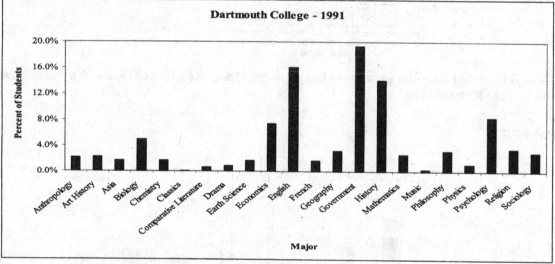

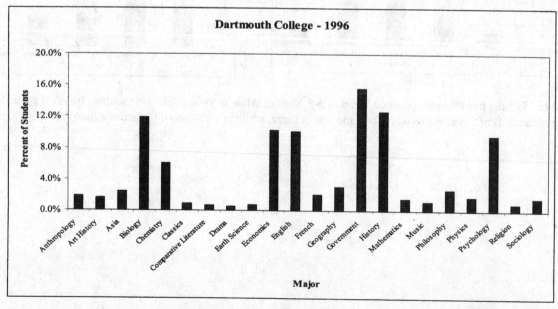

b) Since we used alphabetical order, no changes needed to be made. Descending or ascending order would not be good here.

c) For relative frequency or frequency, it is best to use the same y-scale.

d) Biology and Chemistry have both more than doubled. Physics increased slightly, and Earth Science and Mathematics decreased. Overall, there seems to be an increase.

e)

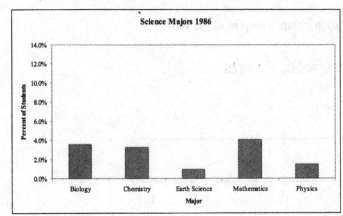 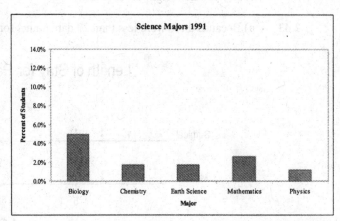

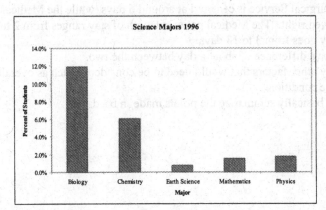

Separate graphs, especially with the same y-scale, do make it easier to focus on the issue.

f) Aside from the Sciences, the Social Studies people are declining, the arts are more or less holding steady, and the Social Sciences are mostly declining a little, except for Psychology.

2.41   a)

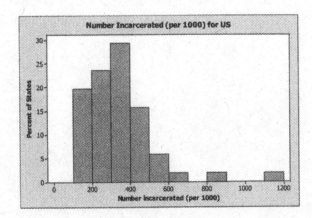

b) Typically there are between 300 and 400 people per 100,000 incarcerated for any state. The data are skewed to the right, and fairly variable. There is an unusual value over between 1100 and 1200.
c) If you follow the guidelines, the histogram will not have enough classes to shown the distribution of the bulk of the data because of the high value.
d) Done right the first time.
e) See part b.
f) No, but did it right the first time.

2.43   a) Because there were less than 25 data points for each group, a dotplot works well.

Length of Stay for Pancreatitis Patients

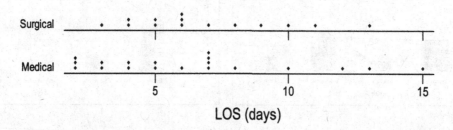

b) The data for the Surgical Service is centered at around 6 days, while the Medical Service is centered at 7 days. Both groups are skewed right. The Medical patients length of stay ranges from 2 to 15days, while the Surgical patients length of stay goes from 3 to 13 days.
c) There appears to be a difference of about a day between the two.
d) No, there are many other factors that would need to be considered such as overall health of the patient, age, gender, and insurance conditions.
e) The memo should basically summarize the points made in b - d.

Chapter 3 Numerical Descriptors of Data - Solutions

Section 3.3 Exercises

3.1   a) $\overline{X}$ = 9.6 min. b) Median = 9.4 min.

3.3   a) $\overline{X}$ = 6.9 hours
b) Median = 6 hours
c) They are not really very different. The data are divided into three sections, low medium and high. The high values skew the mean a little, but either one is a reasonable measure. I like the median a little better for this one, because it's actually a number in the data set, but they really are equivalent.

3.5.   a) $\overline{X}$ = 1.7 million dollars
b) Median = 1.87 million dollars
c) The mean and median are both lower for the LA Clippers. For the Clippers the median is higher than the mean which would indicate a left (lower) skew. For the Celtics the data are skewed to the high side.

3.7   a) $\overline{X}$ = 7.6 Median = 7.5
b) Since the mean and median are virtually equal, the data are probably symmetric.

Section 3.4 Exercises

3.9   a) Range = 9.9
b) s = 3.21; s^2 = 10.31

3.11   a) Range = 4.2; s = 1.37
b) To do this, $\overline{X}$ = 4.6

Percentage	Lower Value	Upper Value
68	4.6 – 1.4 = 3.2	4.6 + 1.4 = 6.0
95	4.6 – (2)(1.4) = 1.8	4.6 + (2)(1.4) = 7.4
>99	4.6 – (3)(1.4) = 0.4	4.6 + (3)(1.4) = 8.8

c) For the data, the percent within one standard deviation is 8/10 = 80%. The percent within two standard deviations is 100%. The actual percentages are high, but that is likely due to the small sample size.

3.13   a) Range = 14 hours; s = 5.3 hours.
b)

Percentage	Lower Value	Upper Value
68%	6.9 – 5.3 = 1.6	6.9 + 5.3 = 12.2
95%	6.9 – (2)(5.3) = -3.7	6.9 + (2)(5.3) = 17.5
>99%	6.9 – (3)(5.3) = -9.0	6.9 + (3)(5.3) = 22.8

c) If you look at the data there are 6/12 values (50%) within one standard deviation of the mean. This would mean that the data are not concentrated around the center, not a nice normal distribution. The Empirical Rule is not a good choice.

d) The z-score is $\frac{4-6.9}{5.3} = -0.55$, which tells me that this observation is almost very slightly below the mean.

3.15   a) Range = 10
b) s = 3.57

c) Since none of the z-scores are more than 2 or less than –2, it would seem that none of the data are unusual.

Data Value	z-score
10	-1.43
12	-0.87
12	-0.87
13	-0.59
14	-0.31
14	-0.31
18	0.81
19	1.09
19	1.09
20	1.37

3.17   a) Range = 3.3; s =1.1
b)

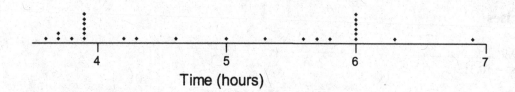

Hours Spent Watching Television

There do not appear to be any unusual values, but there are two distinct centers.
c) The range is probably not too distorted. Both measures do a good job.
d)

Data Value	Z-score
3.6	-1.18
3.7	-1.09
6.3	1.27
6.9	1.82

None of these values are outliers.

Section 3.5 Exercises

3.19   a) $p = \dfrac{8+(1/2)(1)}{12} = 70.8\%$

b) $p = \dfrac{0+(1/2)(1)}{12} = 4.2\%$

c) About 75%

3.21   a) $p = \dfrac{5+(1/2)(1)}{15} = 40\%$

b) $Q_1 = 12$ $Q_3 = 14$

c) It tells them that 50% of their readers make between 12 and 14 subsequent purchases.

3.23 a) $Q_1 = 1$; $Q_3 = 3$ This means that 25% of the rankings were 1 or less and 75% were 3 or less.

b) IQR = 3 - 1 = 2. This tells us that the range of the middle 50% of rankings was 2.

c) The inner fences are at $1 - 1.5 \times 2 = -2$ *and* $3 + 1.5 \times 2 = 6$. No data points fall outside the inner fence, you do not need to find the outer fences.

Food Service Ratings

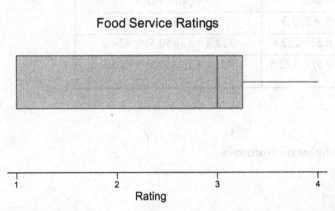

Rating

d) and e) The typical rating was 3. About 50% of the ratings either were between 1 and 3, indicating that half the students were neutral or disliked it. The data are skewed left and the median and third quartile are equal, which really means that 75% of the students were either neutral or disliked it. There were no outliers.

Chapter 3 Exercises
Learning It

3.25 a) $\overline{X} = 13.0$; Median = 13; Mode = 13 and 13.1

b) Range = 0

c) 0.6; s = 0.17

d) v $p = \dfrac{2 + (1/2)(1)}{12} = 20.8\%$

e) $Q_1 = 12.95$; $Q_3 = 13.1$. Fifty percent of the people got relief between 12.95 and 13.1 minutes.

3.27 a) $\overline{X} = 4.84$

b) Median = 4.76

c) The mean and the median are virtually the same, so the data are probably symmetric.

d) No. The mode for continuous data is almost never good. In this case the mode is $4.75, which is close to the center.

e) The range is 0.67.

f) The standard deviation is 0.20.

3.29 a) $\overline{X} = 32.4$; Median = 37.5

b) Range = 29; s = 9.9

c) Since the mean and median differ by more than 10%, and the mean is less than the median, it would appear that the data are skewed left.

Time Spent on Phone

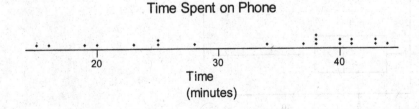

Time
(minutes)

d) This does show a left skew, and a lot of variability.

3.31   a) $\overline{X}$ = 323.8; s = 50.5

b)

Percentage	Lower Value	Upper Value
68	323.8 – 50.5 = 273.3	323.8 + 50.5 = 374.3
95	323.8 – (2)(50.5) =222.8	323.8 + (2)(50.5) = 424.8
>99	323.8 – (3)(50.5) = 172.3	323.8 + (3)(50.5) = 475.3

c)

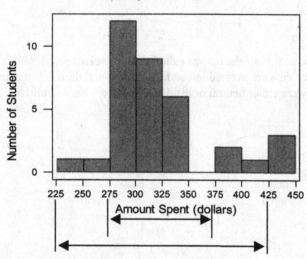

d) 1st Interval = 28/35 = 80%, 2nd Interval = 33/35 = 94.3%, 3rd Interval = 100%

e) The first interval is off by 12%, the second and third are right on. The data are not really symmetric, so the Empirical Rule would not work well here.

3.33   a) Medical: $\overline{X}$ = 6.4; Median = 6; Mode = 7

Surgical: $\overline{X}$ = 6.9; Median = 6; Mode = 6

b) Medical: Range = 13; s = 3.8

Surgical: Range = 10; s = 2.9

c)

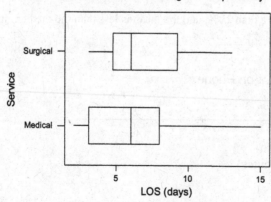

Thinking About It

3.35 **a) No Homework**: A typical student scored between 70 and 80 on the exam. The grades were symmetric with one possibly unusual value. The scores ranged from 60 to 100.
Homework Not Collected: A typical student scored between 70 and 80 on the exam. The grades were skewed to the left a little and variable. They ranged from 40 to 100.

Homework Collected: A typical student scored between 80 and 90. The grades were pretty symmetric, not very variable and ranged from 60 to 100.

There is almost no difference between the group that had no homework assigned and the group that had homework, but did not have it collected. If anything, the second is a little more variable. The group which had homework collected scored about 10 points higher on the average and had less variability. It would seem that collecting homework improved exam scores.

b) Yes and no. There is no proof that the homework was the cause of the grade improvement. If there is no reason to assume that one of the sections was inherently smarter than the others then the experiment probably controlled outside factors. A pre-test of the material might have given information about general differences in the classes. The instructor might want to consider using more than one exam to do the comparison. All things considered, assigning and collecting homework does seem to be effective.

3.37 Since the average changed by 5 points, and the number of items in the sample was 5, the difference in the right number and the wrong number must have been 25. Since the correct value was 10, the incorrect value must have been 35.

3.39 a. Supplier A
$\overline{X}$ = 501.4; Median = 501; Range = 54; s = 10.34
Supplier B
$\overline{X}$ = 495.5; Median = 495; Range = 19;, s = 4.44
b)

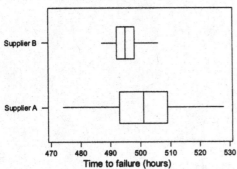

Diskette Failure Times

c) Since longer is better it would appear that Supplier A is the better supplier. Supplier B's disks fail much more quickly than Supplier A's. In fact, 75% of Supplier A's disks fail in less time than 50% of Supplier A. It should be noted that Supplier B's disks are much more consistent. That is, you never see the really short lives that you see with Supplier A, but the times are still much shorter overall.

3.41 The state's average salary is about $150 a month lower, and more than twice as variable. Although they agree on the high end, they have much more variability on the low end. It would seem that they would benefit from moving up the salaries of those people making less than average salaries. This should bring them back in line with industry by increasing the average and decreasing the variability.

3.43   Medical: $\overline{X}$ = 6.4; Median = 6; Mode = 7   Surgical: $\overline{X}$ = 6.9; Median = 6; Mode = 6
Medical: Range = 13; s = 3.8   Surgical: Range = 10; s = 2.9

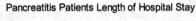

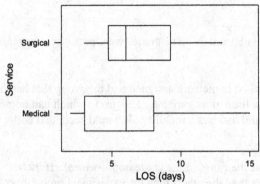

Based on the data it would seem that there is not much difference in the mean or median stay - less than one day. However it is important to note that the medical service has much more variability in the data than the surgical service.
b) No.
c) First of all, there is not that much difference. Second, there are many other factors that would need to be addressed such as age and general health of patient, gender, condition of patient when discharged, insurance requirements and more.
d) The memo should summarize the answers to a - c and include the summary statistics and graph.

3.45   From the data we found that the mean and median crossing times were about 17 seconds. The standard deviation of the times was 3.2 seconds. If the light is set to 15 seconds, you can count on the senior citizens to be in the middle of the street when it changes. If you want to set the light correctly, consider the mean plus 2 or 3 standard deviations - or at least set it to the maximum data value that we obtained.

Chapter 4 Analyzing Bivariate Data - Solutions

Section 4.2 Exercises

4.1　a)

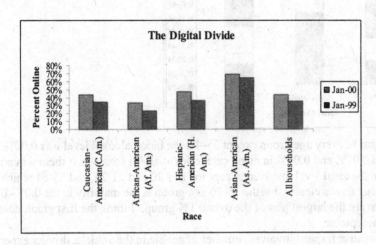

b)

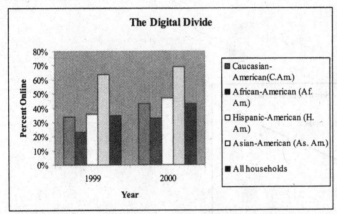

c) Either could be good, depending on the viewpoint. The graph with race on the category axis shows that every group has increased over time, but it is harder to compare races. The graph with year on the category axis shows that over time the percent online has kept about the same pattern with respect to race, that is Asian-American is highest, then Hispanic-Americans, Caucasians and African-Americans.

4.3　a)

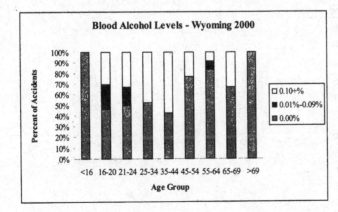

b)

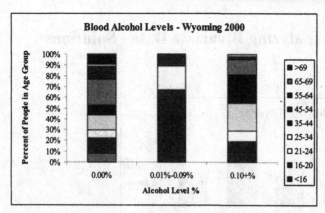

Blood Alcohol Levels - Wyoming 2000

c) From the first graph we see that in every age group except 35-44, the blood alcohol level was 0.00%. Blood alcohol levels were not between 0.01% and 0.09% in many categories at all. It seems as if there was no alcohol involved or there was more than the legal level involved except for 16 - 20 and 21-24, and 55-64 which might be very interesting. The second chart shows clearly that the 16-20 age group is the majority in the 0.01 - 0.09% group. It also shows the two groups that are the largest part of the over 0.1% group. I think the first graph does a slightly better job, but either would be acceptable.

d) Perhaps driving conditions (weather), speed involved, number of people in the vehicle, driving experience of driver (how long licensed?).

Section 4.3 Exercises

4.5   a–c)

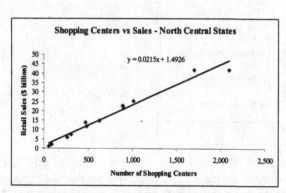

Shopping Centers vs Sales - North Central States

$y = 0.0215x + 1.4926$

This is a very good fit, although it is a bit off on the high values of X.

d)

State	Predicted
Illinois	46.6
Indiana	21.0
Michigan	23.4
Minnesota	11.6
Ohio	38.1
Iowa	8.1
Missouri	20.6
Wisconsin	14.9
South Dakota	2.7
North Dakota	3.4
Nebraska	7.2
Kansas	11.8

4.7 a) and b)

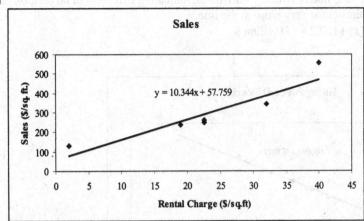

The relationship might be non-linear.

b and c) The intercept indicates that if rental costs were $0, then sales would be $57.76. Why not? Not paying rent does not mean you would not sell anything. The slope says that for every $/sq ft. increase in rent, sales go up $10.

c) The line does an acceptable job at predicting, though there are some big discrepancies, notably Jewelry and Food Service.

Chapter 4 Exercises
Learning It

4.9 a)

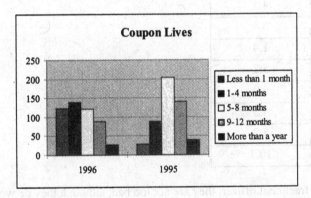

b) In 1995, the highest category was 5-8 months. In 1996 it is 1-4 months. From 1995 to 1996 the lower categories (less than one month and 1-4 months) grow considerably.

4.11 a) and b)

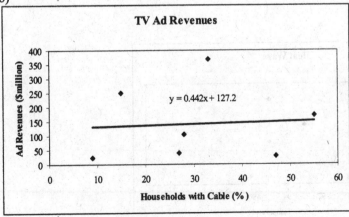

c) The line does a terrible job. The line is virtually horizontal, indicating that there in no relationship between the two variables. None of the points come very close to the line.

d) For the Ukraine: $\hat{y} = 0.442(x) + 127.2 = 131$ billion $.

4.13   a)–c)

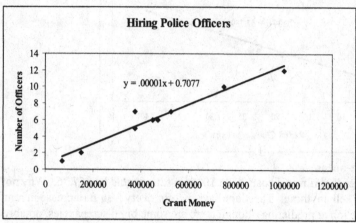

d) This appears to be a really good fit.

e)

Grant $	# Officers	Predicted #
150000	2	2.2
375000	5	4.5
471125	6	5.4
70967	1	1.4
450000	6	5.2
525000	7	6.0
375370	7	4.5
750000	10	8.2
1000000	12	10.7

f) If you consider that you can't hire a fractional officer, they are not too bad, although they do worse at the higher levels.

Thinking About It

4.15   a) I think that this is likely, yes.
            b)

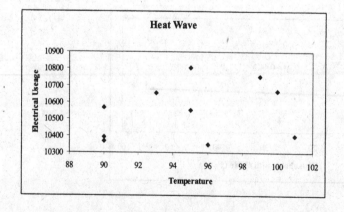

c) There doesn't seem to be any relationship at all here. The points seem to be scattered randomly with no pattern.
d) The inclusion of data from different years might cause a problem.
e)

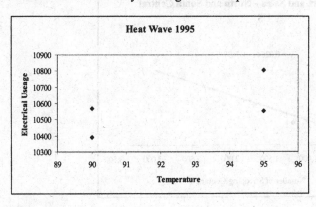

It doesn't really help much.
f) The data point for 7/9/95 is rather unusual. Not without a calendar.
g) and h)

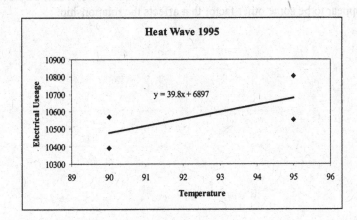

Now there is more of a relationship. There is still a lot of variability, but the line makes sense.

4.17   a)

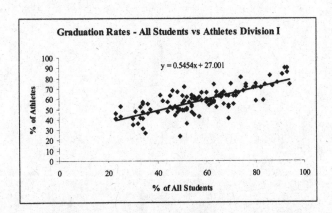

b) The LS line slope indicates that for every 1% increase in overall graduation rate, the rate for athletes goes up by 0.5%. The intercept says that if the % of students that graduate is 0%, the percent of athletes will be 27%, which makes no sense at all since you can't have athletes without students.
c) The line fit is ok - not great. There is clearly a relationship, but many points are far off the line, particularly for low athlete graduation rates.

4.19   a) and b)

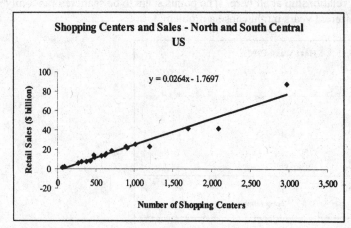

c) The answer could go either way. The lines were similar, and the fit is still very good (maybe better) for the lower X values. Even adding the higher X value states (Texas, Tennessee, Illinois, Ohio) the fit is still bad at the high end. I would be opposed to the practice of combining as a rule, since if there are differences, they will be masked by it. In this case there does not appear to be some other factor that affects the relationship.

Chapter 5 Probability - Solutions

Section 5.2 Exercises

5.1   a) 559 / 1017
       b) 458 / 1017

5.3   a) 280 / 450
       b) 170 / 450
       c) They are estimates. The 450 students are not the whole population.

5.5   a) 740 / 6492
       b) 3427 / 6492
       c) 258 / 6492
       d) Not from Brookfield = 6492 - 302 = 6190, so 6190 / 6492

Section 5.3 Exercises

5.7   a) 284 / 601
       b) 13 / 601
       c) (239 + 39) / 601 = 278 / 601
       d) (300 + 284 - 125) / 601 = 459 / 601

5.9

Frequency of Taking Medicine	Type of Medication		Totals
	Prescription	**Over-the-counter**	
Daily	86	43	129
As Needed (sporadic)	23	156	179
Totals	109	199	308

a) $\dfrac{129}{308}$

b) $\dfrac{199 + 129 - 43}{308} = \dfrac{285}{308}$

c) $\dfrac{43}{308}$

Section 5.4 Exercises

5.11   a) 125 / 300
        b) 13 / 39
        c) 263 / 301
        d) P(Female) = 301 / 601 P(Not Sure or Didn't Know) = 39 / 601 ; P(Female AND NS/DK) = 25 / 601

If we multiply $\dfrac{301}{601} \times \dfrac{39}{601} = 0.032$ and compare it to $\dfrac{25}{601} = 0.042$ it would appear that they are not independent in this sample. They differ by about 1%.

5.13

Frequency of Taking Medicine	Type of Medication		
	Prescription	Over-the-counter	Totals
Daily	86	43	129
As Needed (sporadic)	23	156	179
Totals	109	199	308

a) $\dfrac{43}{199}$

b) $\dfrac{86}{109}$

c) $\dfrac{156}{199}$

d) P(over the counter) = 199/308 and P(sporadically)=179/308. P(over the counter AND sporadically) = 156/308 = 0.506. Since $\dfrac{199}{308}*\dfrac{179}{308} = 0.375$ it would appear that the events are not independent in this sample.

Chapter 5 Exercises
Learning It

5.15   a) 1105 / 1730 = 0.639
        b) 160 / 1730 = 0.092

5.17   a) 868 / 1000 = 0.868
        b) This poll gives the percentage as about 23% higher than the original Gallup poll in 1953.

5.19   a) 436 / 2200
        b) 231 / 2200
        c) 1 - (1125 / 2200) = 1075 / 2200, or alternatively, (436 + 231 + 215 + 193) / 2200 = 1075 / 2200
        d) That leaves reading. 231 / 2200

5.21

Number of Times Had Five or More Drinks in a Row in Last Two Weeks

GENDER	0	1	2	3	4	5	Total
Male	55	20	18	19	8	4	124
Female	75	14	12	24	2	0	127
Total	130	34	30	43	10	4	251

a) P(Male and 2 times) = $\dfrac{18}{251}$

b) P(Female or 3 times) = $\dfrac{127+43-24}{251}=\dfrac{146}{251}$

c) P(3 or 4 times) = $\dfrac{43+10}{251}=\dfrac{53}{251}$

d) P(0 times) = $\dfrac{130}{251}$

Thinking About It

5.23   a) Considering the state of the world at both times it might not be reasonable because in 2002 the US was still reeling from the attack on the World Trade Center and there was a definite change in peoples' attitudes toward spirituality. Also, the second poll came on the heels of the court decision which would catch people's attention. The 1953 poll was conducted as the US was considering adding the words - the sentimentalism might not have been as strong.

b) The new poll might be biased for the reasons stated in part a).

c) If the sample were truly random it would be representative of Americans at that time - maybe not in another time.

5.25   a)

	Attempted suicide	Did not attempt suicide	Total
Adopted	16	198	214
Not Adopted	197	6166	6363
Total	213	6364	6577

b) P(Suicide/Adopted) = 16 / 214 = 0.075

c) P(Suicide/ Not Adopted) = 197 / 6363 = 0.031

d) From the answers to b) and c) it would appear that the rate is twice as high for adopted children. If we check the probabilities: P(Adopted) = 214 / 6577 , P(Suicide) = 213 / 6577, P(Adopted AND Suicide) = 16 / 6577. Now,

$$\frac{214}{6577} \times \frac{213}{6577} = \frac{45582}{43256929} = 0.001 \text{ and } \frac{16}{6577} = 0.002 \text{ so they are not independent.}$$

5.27   a) 91 / 281

b) 8 / 13

c) For everyone the percent at 0.10+ is 91 / 281 (32.4%) and for those with DWI convictions it is 8 / 13 (61.5%), almost double.

d) 4 / 13

e) It would appear that people with previous DWI convictions are more likely to have very high blood alcohol levels.

5.29

		Hangover Since Beginning of Semester		
GENDER	**Not At All**	**Once**	**Twice or More**	**Total**
Male	61	23	40	124
Female	66	25	36	127
Total	127	48	76	251

a) P(Twice or more/female) = 36 / 127

b) P(Male / Once or Less) = (61 + 23) / (127 + 48) = 64 / 175

c) P(Two or more/male) = 40 / 124

d) For females the probability of being hung over twice or more is 36 / 127, (28.3%), while for males it is 40 / 124 or 32.3%. For the males it is slightly higher but not amazingly so. This would mean that males are not much more likely than females to be hung over twice or more a semester.

Chapter 6 Random Variables and Probability Distributions - Solutions

Section 6.2 Exercises

6.1

x	0	1	2	3	4	5	6
p(x)	0.10	0.23	0.18	0.16	0.13	0.10	0.10

a) $P(X \le 4) = 0.10 + 0.23 + 0.18 + 0.16 + 0.13 = 0.80$
b) $P(2 < X < 4) = P(X = 3) = 0.16$
c) $P(X > 4) = 0.10 + 0.10 = 0.20$

6.3

x	0	1	2	3	4
p(x)	0.23	0.34	0.17	0.15	?

a) Since the probabilities must add to 1 and the others total 0.89, $p(4) = 0.11$
b) $(X = 0) = 0.23$
c) $P(X \ge 1) = 0.34 + 0.17 + 0.15 + 0.11 = 0.77$
d) $P(X > 2) = 0.15 + 0.11 = 0.26$

6.5

x	0	1	2	3	4	5	6	7	8
p(x)	0.05	0.08	0.13	0.23	0.18	0.13	0.08	0.06	0.06

a) $P(X \ge 3) = 0.23 + 0.18 + 0.13 + 0.08 + 0.06 + 0.06 = 0.74$
b) $P(2 < X < 5) = P(X = 3 \text{ or } 4) = 0.23 + 0.18 = 0.41$
c) Not more than 6 means 6 or less so, $P(X \le 6) = 0.05 + 0.08 + \ldots + 0.08 = 0.88$
d) If they sold 39 tickets then at least 4 people have to not show for there to be seats for everyone. So, $P(X \ge 4) = 0.18 + \ldots + 0.06 = 0.51$

Section 6.3 Exercises

6.7 For this problem, n = 20 and p = 0.25
a) $P(X \le 5) = 0.003 + 0.021 + \ldots + 0.202 = 0.617$
b) $P(4 \le X \le 8) = P(X = 4, 5, 6, 7 \text{ or } 8) = 0.190 + 0.202 + 0.169 + 0.112 + 0.061 = 0.734$
c) $P(X > 7) = 0.061 + 0.027 + \ldots + 0 = 0.102$
d) Now, p changes to 0.75 because we are talking about the tickets that HAVE been paid. $P(X \ge 6) = 0 + 0 + 0.001 + \ldots + 0.003 = 1.00$

6.9 For this problem, n = 25 and p = 0.70 or 0.30

a) (p = 0.30) $P(X \le 5) = 0.000 + \ldots + 0.103 = 0.192$
b) (p = 0.70) $P(10 < X < 17) = 0.004 + \ldots + 0.134 = 0.322$
c) $\mu = 25 * 0.70 = 17.5$
d) $\sigma = \sqrt{25 * 0.70 * 0.30} = 2.29$
e) $\mu \pm 3\sigma = 17.5 + 3 * 2.29 = 24.37$
   $17.5 - 3 * 2.29 = 10.63$ so $P(10.63 < X < 24.37) = P(11 \le X \le 24) = 0.997$.

Section 6.5 Exercises

6.11   X~N(235,15)

    a)

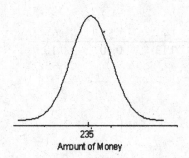

235
Amount of Money

    b)

Z score	Table Value
$Z = \dfrac{220 - 235}{15} = -1.00$	0.1587
$Z = \dfrac{250 - 235}{15} = 1.00$	0.8413

So P(220 < X < 250) = 0.8413 - 0.1587 = 0.6826

    c)

Z score	Table Value
$Z = \dfrac{270 - 235}{15} = 2.33$	0.9901

P(X > 270) = 1 - 0.9901 = 0.0099 or 0.99%

    d)

Z score	Table Value
$Z = \dfrac{225 - 235}{15} = -0.67$	0.2514

P(X < 225) = 0.2514 0r 25.14%

6.13   X~N(65, 0.35)

    a)

Z score	Table Value
$Z = \dfrac{64.5 - 65}{0.35} = -1.43$	0.0.0764

P(X > 64.5) = 1 − 0.0764 = 0.9236 or 92.36%

    b)

Z score	Table Value
$Z = \dfrac{64 - 65}{0.35} = -2.86$	0.0021
$Z = \dfrac{66 - 65}{0.35} = 2.86$	0.9979

So P(64 < X < 66) = 0.9979 - 0.0021 = 0.9958 or 99.58%

c) What percentage is actually above the labeled weight (64 oz.)?

Z score	Table Value
$Z = \dfrac{64-65}{0.35} = -2.86$	0.0021

P(X > 64) = 1 - 0.0021 = 0.9979 or 99.79%, so the company meets the requirements.

Chapter 6 Exercises
Learning It

6.15 Binomial with n = 15 and p = 0.30 or 0.70
a) P(X ≥ 5) = 0.206 + ... + 0.000 = 0.485
b) P(6 ≤ X ≤ 12) = 0.279
c) p = 0.70; P(X < 10) = p(0) + ... + p(9) = 0.279

6.17 Binomial with n = 20 and p = 0.20 or 0.80
a) P(X < 8) = p(0) + ... + p(7) = 0.969
b p = 0.80; P(X ≤ 10) = 0.002
c) p = 0.20; P(7 < X < 19) = p(8) + ... + p(18) = 0.031
d) $\mu = 20 \times 0.20 = 4$; $\sigma = \sqrt{20 \times 0.20 \times 0.80} = 1.79$

6.19 X ~ N(350, 28)
a)

Z score	Table Value
$Z = \dfrac{300-350}{28} = -1.79$	0.0367
$Z = \dfrac{400-350}{28} = 1.79$	0.9633

P(300 < X < 400) = 0.9633 - 0.0367 = 0.9266 = 92.66%

b)

Z score	Table Value
$Z = \dfrac{375-350}{28} = 0.89$	0.8133

P(X > 375) = 1 - 0.8133 = 0.1867

c)

Z score	Table Value
$Z = \dfrac{425-350}{28} = 2.68$	0.9963

P(X ≤ 425) = 0.9963

6.21 Binomial with n = 15 and p = 0.30 or 0.70
a) P(4 ≤ X ≤ 7) = p(4) + ... + p(7) = 0.653
b) P(X < 6) = p(0) + ... + p(5) = 0.721
c) P(X ≥ 9) = p(9) + ... + p(15) = 0.015
d) p = 0.70 P(5 < x < 10) = p(6) + ... + p(9) = 0.275

6.23   X~N(90,6)

a)

Z score	Table Value
$\frac{105-90}{6} = 2.50$	0.9938

P(X > 105) = 1 - .9938 = 0.0062 = 0.62%

b)

Z score	Table Value
$Z = \frac{85-90}{6} = -0.83$	0.2033
$Z = \frac{105-90}{6} = 2.5$	0.9938

So P(85 < X < 155) = 0.9938 - 0.2033 = 0.7905 or 79.05%

c)

Z score	Table Value
$Z = \frac{110-90}{6} = 3.33$	0.9996

P(X < 110) = 0.9996 or 99.96%

6.25   Binomial with n = 20 and p = 0.60 or 0.40
a) P(X ≥ 15) = p(15) + ... + p(20) = 0.126
b) P(12 < X < 17) = p(13) + ... + p(16) = 0.400
c) P(X < 5) = p(0) + ... + p(4) = 0.051
d) P(X ≤ 13) = p(0) + ... + p(13) = 0.994

Thinking About It

6.27   a) μ =(25)(0.90) = 22.5
b) $\sigma = \sqrt{25 \times 0.90 \times 0.10} = 1.5$
c) We are looking for P(μ − 2σ < X < μ + 2σ): μ − 2σ = 22.5 - 2(1.5) = 19.5 and   μ + 2σ = 22.5 + 2(1.5) = 25.5
So, P(19.5 < X < 25.5) = P(20 ≤ X ≤ 25) = p(20) + ... + p(25) = 0.966
d) I would not really expect it to agree as well as it did because the binomial with p = 0.90 is not symmetric, but it does compare well. It is a little high.

6.29   a) From the graph it would appear that the hourly pay for the clean houses is higher and less variable than for the not clean houses. There is some overlap between the two populations.

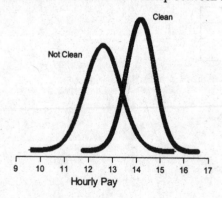

b) The distributions for years of education look amazingly similar to those for hourly pay. There might be a little more overlap between the distributions.

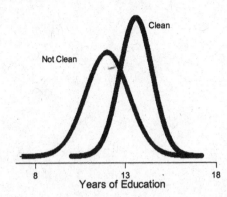

c) For clean houses X~N(13.6, 0.9):

Z score	Table Value
$\frac{16-13.6}{0.9} = 2.67$	0.9962

P(X ≥ 16) = 1 - 0.9962 = 0.0038

d) For not clean homes X~N(12, 1.2)

Z score	Table Value
$\frac{16-12}{1.2} = 3.33$	0.9996

P(X ≥ 16) = 1 - 0.9996 = 0.0004

e) For both groups the percent who finish college is less than 1%. However, for the clean home groups, the mean is 13.6 which indicates some college and 50% will have more than that. For the not clean homes the mean is just graduated high school.

6.31 a) The z score that has 10% of the area above it (or 90% of the area below it) is 1.28 (area = 0.8997). So we are looking for a value that is 1.28 standard deviations above the mean.
The value, X, is 3.4 + 1.28(0.9) = 4.55 hours. So to get into the "very clean" category you have to clean your home at least 4.55 hours per week.
b) The z score that has 2% of the data below it is –2.05 (area = 0.0202), so we are looking for a value that is 2.05 standard deviations below the mean.
The value is 3.4 – 2.05(0.9) = 1.55 hours per week to be "uninhabitable".

6.33 a) It is unusual. It is not within 2 standard deviations of the mean. In fact, it is not even within 3 standard deviations of the mean.
b) It might mean that they had strange sample, or it might mean that the belief that the percentage is 60% might be wrong.

6.35 The norm for missing class because of drinking is 60%. The percentage that the dean found is 15/20 or 75%. It seems a bit high. If we calculate expected number we would find it to be $\mu = 12$. The standard deviation is $\sigma = 2.19$. If you go up 2 standard deviations (or about 4) you are at 12 + 4 = 16. The 15 students she found it not all that unusual. I would agree with the dean's assessment, although I would be horrified.

Chapter 7 Sampling Distributions and Confidence Intervals -Solutions

Section 7.3 Exercises

7.1   a) $\overline{X}_{Sosa} = 407.5$

       b) $\overline{X}_{McGwire} = 402.3$

       c) $\overline{X}_{Sosa} - \overline{X}_{McGwore} = 5.2$

       d) The difference is called the sampling error.

7.3   a) $\overline{X} = 6.3$ minutes

       b) 3.7 minutes

7.5   a) $\overline{X} = 118.92$ minutes

       b) Yes. The difference is $118.92 - 120 = -1.08$ minutes

Section 7.7 Exercises

7.7   a) $\overline{X} = 0.17$ inches

       b) $\sigma_{\overline{x}} = \dfrac{0.1}{\sqrt{30}} = 0.018$

       c) Since there are 31 days in May, $4.88 / 31 = 0.16$ inches.

       d) $Z = \dfrac{0.17 - 0.16}{0.018} = 0.56$ This seems to indicate that this May was not unusual.

7.9 $\overline{X} = 80$ and $\sigma_{\overline{x}} = \dfrac{5}{\sqrt{50}} = 0.71$; $Z = \dfrac{80-75}{0.71} = 7.04$

       Yes, the high school can be considered "unusually" strong.

Section 7.8 Exercises

7.11   a) $\overline{X} = 382$; $n = 45$; $\sigma = 21$

       $\sigma_{\overline{x}} = \dfrac{21}{\sqrt{45}} = 3.13$

       Error $= 2.33 * 3.13 = 7.29$

       Lower: $382 - 7.29 = 374.71$

       Upper: $382 + 7.29 = 389.29$

       b) We are 98% confident that the true mean is between 374.71 and 389.29.

7.13   a) $\overline{X} = 22.7$; $n = 53$; $\sigma = 1.5$

       $\sigma_{\overline{x}} = \dfrac{1.5}{\sqrt{53}} = 0.21$

       Error $= 1.96 \times 0.21 = 0.41$

       Lower: $22.7 - 0.41 = 22.29$

       Upper: $22.7 + 0.41 = 23.11$

       b) $\overline{X} = 29.6$; $n = 44$; $\sigma = 1.8$

       $\sigma_{\overline{x}} = \dfrac{1.8}{\sqrt{44}} = 0.27$

       Error $= 1.96 \times 0.27 = 0.53$

       Lower: $29.6 - 0.53 = 29.07$

       Upper: $29.6 + 0.53 = 30.13$

7.15   **One-Sample Z: All Years**
The assumed sigma = 0.15

```
Variable          N      Mean    StDev    SE Mean       90.0% CI
All Years        120    3.0691   0.1560   0.0137    ( 3.0465,  3.0916)
```

7.17   **Note: We are using s instead of sigma since n is large.**
a) **One-Sample Z: Shaquille**
The assumed sigma = 9.04

```
Variable          N      Mean    StDev    SE Mean       95.0% CI
Shaquille        40     25.55     9.04     1.43     ( 22.75,   28.35)
```
b) He was asleep? Actually he probably didn't play much of the game…or he didn't play at all.
c) **One-Sample Z: Shaquille**
The assumed sigma = 7

```
Variable          N      Mean    StDev    SE Mean       95.0% CI
Shaquille        38     26.89     7.00     1.14     ( 24.67,   29.12)
```
d) From the confidence interval we expect that 95% of the time the interval from 24.67 to 29.12 points will contain Shaq's true average. Since 25 is in the interval, he is not really that different.

7.19   a) $\overline{X} = 3.12$; $n = 50$; $s = 1.04$

$\sigma_{\bar{x}} = 1.04 / \sqrt{50} = 0.147$

Error = $1.645*0.147 = 0.242$
Lower = $3.12 - 0.242 = 2.88$
Upper = $3.12 + 0.242 = 3.36$

b) The management of the symphony can be 90% confident that the average number of times a person would attend a concert is between 2.88 and 3.36, or, for all practical purposes, close to 3.

Section 7.10 Exercises

7.21   a) $\overline{X} = 62.4$; $n = 10$; $s = 11.82$
$\alpha = 0.05$; df = 9; Use $t_{0.025,9} = 2.262$

Error = $2.262 * \left(11.82 / \sqrt{10} \right) = 8.45$

Lower = $62.4 - 8.45 = 53.95$
Upper = $62.4 + 8.45 = 70.85$

b) Yes, the park should increase staff. The confidence interval suggests that the true average wait time is likely to be as high as 71 minutes.

7.23 $\overline{X} = 2.52$; $n = 10$; $s = 0.16$
$\alpha = 0.05$ so use $t_{0.025,9} = 2.262$

Error = $2.262 \times \left(0.16 / \sqrt{10} \right) = 0.114$

Lower = $2.52 - 0.114 = 2.41$
Upper = $2.52 + 0.114 = 2.63$

Section 7.11 Exercises

7.25   a) $p = 0.35$; $n = 100$; 95% Confidence

Error = $1.96 * \sqrt{\dfrac{(0.35)(0.65)}{100}} = 0.093$

Lower = 0.35 − 0.093 = 0.257
Upper = 0.35 + 0.093 = .443
b) Since the target value of 0.35 is in the interval, they will not cancel the show.

7.27   a) p = 0.46; n = 450; 95% Confidence

$$Error = 1.96*\sqrt{\frac{(0.46)(0.54)}{450}} = 0.046$$

Lower = 0.46 − 0.046 = 0.414
Upper = 0.46 + 0.046 = 0.506

b) Yes, it is likely that 50% of the Bay State Voters are opposed to allowing casino gambling in the state.
c) If the governor wants to have his own way, he will look at the fact that the percentage of people who oppose the gambling is likely to be as low as 41.4% which is clearly a minority. He might push it through.

7.29   a) p = 0.33; n = 100; 95% Confidence

$$Error = 1.96*\sqrt{\frac{(0.33)(0.67)}{100}} = 0.092$$

Lower = 0.33 − 0.092 = 0.238
Upper = 0.33 + 0.092 = 0.422
Since it appears that the percentage of students who study in their rooms is between 24% and 42%, the university should plan for 0.33% which is the point estimate.

Section 7.12 Exercises

7.31   Z = 1.96; σ = 1.5; e = 0.25

$$\frac{1.96^2 \times 1.5^2}{0.25^2} = 138.30 \text{, so } 139$$

7.33   Z = 1.645; σ = 165; e =10.

$$n = \frac{(1.645^2)(165^2)}{10^2} = 736.72, \quad so \quad 737$$

7.35   Again, use p = 0.50 to get e = 0.02 with 95% confidence

$$n = \frac{(1.96^2)(.5)(1-0.5)}{.02^2} = 2401$$

Chapter 7 Exercises
Learning It

7.37   a) $\overline{X}$ = 118.92; s = 1.41
b) Since n = 25 and we do not know σ, we use t. With α = 0.05, $t_{0.025,24} = 2.064$.

$$Error = 2.064*\left(\frac{1.41}{\sqrt{25}}\right) = 0.582 \text{ and}$$

Lower = 118.92 − 0.582 = 118.34
Upper = 118.92 + 0.582 = 119.50
c) The claim would appear false, since we are 95% sure that the true mean time is between 118.34 and 119.50 and 120 is not in this interval.

7.39   a) $\overline{X} = 39.425$; $s = 5.773$

b) The standard error is $5.773\big/\sqrt{40} = 0.913$.

c) This is included in the output just next to the value of the standard deviation.

d) $\dfrac{39.425 - 30}{0.913} = 10.32$

e) Based on this value, the assumption of 30 seconds is unreasonable.

7.41 $\overline{X} = 0.92$; $\sigma_{\bar{x}} = 0.05\big/\sqrt{36} = 0.009$

Error = $1.645 * 0.009 = 0.015$

Lower = $0.92 - 0.015 = 0.905$

Upper = $0.92 + 0.015 = 0.935$

7.43   Error = 0.75; $\sigma = 5$

a) 99%; Z = 2.326

$n = \dfrac{(2.326^2)(5^2)}{0.75^2} = 240.5,\quad so\quad 241$

b) 95%; Z = 1.96

$n = \dfrac{(1.96^2)(5^2)}{0.75^2} = 170.7,\quad so\quad 171$

c) The sample size for the 95% confidence interval is smaller. This makes sense since it should cost more to obtain a better estimate.

Thinking About It

7.45   a) $X \sim N(8, 2.5)$

b) $\overline{X} \sim N(8, 0.56)$

c) $\overline{X} \sim N(8, 0.46)$

d) $\overline{X} \sim N(8, 0.28)$

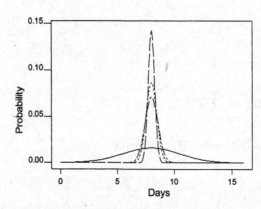

e) As the sample size increases, the width of the sampling distribution decreases. The estimate improves.

7.47   a) They are all in the northeast.

b) This data would probably not apply to all states in all regions of the country.

c) We need to assume that the population (all states) is normally distributed.

One-Sample T: States

Variable	N	Mean	StDev	SE Mean	90.0% CI
States	6	5270675	7047433	2877102	(-528935,11070286)

d) You have to consider the size of the state as well. Population density (people per square mile or acre) would be better.

7.49   Since the margin of error is 5%, this means that the true percent could be as low as 42% or as high as 52%. It is therefore possible that less than half of the population agrees with him.

Chapter 8 Hypothesis Testing: An Introduction - Solutions

Note: A TI-83 calculator was used for the results of many hypothesis test questions. If you use a calculator and tables you might get slightly different answers due to rounding.

Section 8.6 Exercises

8.1   a)–c) $\overline{X} = 195.1$

H_o:	$\mu = 130.5$
H_A:	$\mu \neq 130.5$
Critical Values	± 1.96
Test Statistic	1.35
p value	0.176

d) There is definitely one outlier, and the data look a bit skewed.

Dotplot of Suspensions

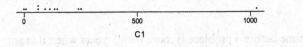

C1

e) Since the p value is larger than 0.05, we cannot reject H_o. There is not enough evidence to say that the average number of suspensions has changed.

8.3   a)–c) $\overline{X} = 7.82$

H_o:	$\mu = 7.8$
H_A:	$\mu \neq 7.8$
Critical Values	± 1.96
Test Statistic	1.48
p value	0.1400

d) Since p is greater than 0.05, we cannot reject H_o. There is not enough evidence to say that the average amount dispensed is different from 7.8 ounces.

8.5   a)–c)

H_o:	$\mu = 60$
H_A:	$\mu \neq 60$
Critical Values	± 1.96
Test Statistic	-5.916
p value	0.000

d) Since the p value is less than 0.05, we reject H_o. The mean time spent playing video games is not 60 minutes.

Section 8.7 Exercises

8.7 H_o: $\pi \leq 0.25$
H_A: $\pi > 0.25$

a) Type I Error: Say the proportion is greater than 25% when it is not. That is, they would hold the orientation session when they should not.

b) Type II error: Do not say the proportion is greater than 25% when it is. That is, do not have the orientation session when they should.

c) The Type I error is more costly, so they should test at 0.05 or 0.01.

8.9 H_o: $\mu \geq 1$

H_A: $\mu < 1$

a) Type I Error: Say the average healing time is less than 1 week when it is not.

b) Type II Error: Do not say the healing time is less than 1 week when in fact it is.

c) I can't see either error as earth shattering, so you can probably test at 0.05.

8.11   *Assuming you want to prove the NHTSA correct:*

H_o: The number of fatal crashes for ABS is at least as great as without

H_A: The number of fatal crashes for ABS is less than no ABS.

a) Type I: Say the mean for ABS is less than no ABS, when it is not.

b) Type II: Do not say the mean for ABS is less than no ABS when it is.

c) It really depends on the consequences of the decision. If the NHTSA is going to force manufacturers to use ABS brakes, then the type II error is more serious, because more lives would be saved. On the other hand, from an auto manufacturer's point of view, if ABS brakes are costly, the Type I error is worse. In the interest of safety, I would say to test at 0.10 so that the Type II error is reduced.

8.13 H_o: $\mu \leq 3$

H_A: $\mu > 3$

a) Type I: Conclude that the average time before a problem is more than 3 years when it is not.

b) Type II: Do not conclude that the mean time is more than 3 years when it is.

c) This will require a cost analysis. If you make a Type I error, you will pay for the new warrenty when you don't need to. If you make a Type II error, then it will depend on the average cost of a repair compared to the warranty cost. I would suggest 0.05.

8.15 H_o: $\mu \leq 1$

H_A: $\mu > 1$

a) Type I: Say the mean is greater than 1 hour when it is not. (Buy more computers when you don't need them).

b) Type II: Do not say the mean is greater than 1 hour when it is. (Don't buy more computers when they are justified).

c) Since computers are expensive, the Type I error is worse in the short run. If you look long range, the ramifications of a Type II error might be loss of students, but that will probably not happen too quickly. Testing at 0.01 is probably best.

Section 8.8 Exercises

8.17 H_o: $\pi \leq 0.80$

H_A: $\pi > 0.80$

8.19 H_o: $\mu \geq 1$

H_A: $\mu < 1$

8.21   *Assuming you want to prove the NHTSA correct:*

H_o: The number of fatal crashes for ABS is at least as great as without

H_A: The number of fatal crashes for ABS is less.

8.23 H_o: $\mu_C = \mu_S$

H_A: $\mu_C \neq \mu_S$

8.25 H_o: $\mu_s \leq 25$
            H_A: $\mu_s > 25$

8.27 H_o: $\mu \leq 1$
            H_A: $\mu > 1$

Section 8.9 Exercises

8.29   a)–c)

H_o:	$\mu \leq 18$
H_A:	$\mu > 18$
Critical Values	1.28
Test Statistic	3.550
p value	0.0002

d) Since the p value of 0.0001 is less than 0.10, we reject H_o and conclude that the mean life is greater than 18 months.

8.31   a)–c) $\overline{X} = 115.47$

H_o:	$\mu \leq 115$
H_A:	$\mu > 115$
Critical Values	2.33
Test Statistic	0.733
p value	0.2317

d) Since the p value is greater than 0.01, we cannot reject H_o. There is not enough evidence to say that the average service speed has increased. The tennis pro should not invest in the new racket.

Chapter 8 Exercises

Learning It
8.33   a) As stated, the claim is that the average time is 3.5 minutes, which is a two sided claim.
            b) H_o: $\mu = 3.5$
               H_A: $\mu \neq 3.5$
            c) Type I: Say the average time is not 3.5 when it is. (Don't let them make the claim when they are correct)
            d) Type II: Don't say the average is not 3.5, when it is not. (Let them make the claim when they are wrong).
            e)

H_o:	$\mu = 3.5$
H_A:	$\mu \neq 3.5$
Critical Values	± 2.326
Test Statistic	-2.530
p value	0.0114

Since the p value is less than the level of significance, we reject H_o and say that the mean is not 3.5 They cannot make the claim on television.

Note: Had they done a one–sided test with
H_o: $\mu \geq 3.5$
H_A: $\mu < 3.5$

They would have rejected H_o and said that the mean is less than 3.5 minutes (even better). However, with the hypotheses set up this way, they might have rejected H_o if the mean were really exactly 3.5.

8.35   a) One sided (because it says reduced)

b) H_0: $\mu \geq 68$
   H_A: $\mu < 68$

c) Type I: Say the mean is less than 68 mph when it is not.

d) Type II: Do not say the mean is less than 68 mph, when it is.

e) and f)

H_0:	$\mu \geq 68$
H_A:	$\mu < 68$
Test Statistic	-17.321
p value	0.0000

Since the p value is clearly zero, no matter what the level of significance, we reject H_0. The mean speed is less than 68 mph.

8.37   a) Two tailed – we simply want to know if it is atypical – either longer or shorter.

b) H_0: $\mu = 55$
   H_A: $\mu \neq 55$

c) Type I: Say the wait is not 55 days when it is.

d) Type II: Say the wait is 55 when it is not.

e) and f)

H_0:	$\mu = 55$
H_A:	$\mu \neq 55$
Critical Values	± 1.96
Test Statistic	7.825
p value	0.0000

Since the p value is less than the level of significance, we reject H_0. The mean is not 55 days. If we redid the problem as a one sided test to say that it is significantly more than 55 days, then I would be looking for a different hospital.

Thinking About It

8.39   It is a two-sided test, and you can't switch the hypotheses for a two sided test. It would be possible to redo the test as a one-sided test if you had reason to.

8.41   a) If your null hypothesis were H_0: $\mu \geq 18$, then you would be trying to prove that the new drug actually decreases life expectancy.

b) You would have to consider general health of the patient, age, time with the disease and many other factors.

8.43 H_0: $\mu = 20000$
H_A: $\mu \neq 20000$
Type I Error: Say tourism is not 20000 when it is.
Type II Error: Say tourism is 20000 when it is not.

a) From the governments perspective I can't see either one being more critical than the other. From a hotel owner's perspective, I would think that the Type II error is worse. An error in estimate might make me plan for the wrong number of clients which would negatively impact my service. If I get less customers than I expect I lose revenue and if I get more I might be short on supplies and help.

b) Since the p value is greater than 0.10, we cannot reject H_0. There is not enough evidence to say that the mean number of tourists per month is different from 20000

8.45 H_o: The drug is safe

H_A: The drug is not safe

a) Type I: Say the drug is not safe when it is.

b) Type II: Do not say the drug is not safe when it is not safe.

c) They are exactly reversed.

d) The first setup is better. In that method, you prove the drug is safe if you reject H_o. In this setup, when you fail to reject H_o, you simply say that there is not enough evidence to say it is not safe…you NEVER say it is safe.

8.47   If they make a Type I error, they will stop the process when they should not. They lose production time, hunting for problems that do not exist, or make changes to a process that did not need them. If they make a Type II error, they will let the process run in error. If they check regularly, they might make items that are defective, but the next check should pick up the problem.

8.49   The Type I error, opening the restaurant when you should not, is more costly. Not only do you have the costs associated with opening a restaurant, but you do not have the profits you need. The Type II error means you don't open a restaurant where there might be adequate business, but there are other places to open restaurants, so you don't lose entirely.

Chapter 9 Inferences: One Population -Solutions

Note: A TI-83 calculator was used for the results of many hypothesis test questions. If you use a calculator and tables you might get slightly different answers due to rounding.

Section 9.2 Exercises

9.1 a) This is a t test. Assume a level of significance of 0.05. $\overline{X} = 12.38$ and $s = 13.55$.

H_0:	$\mu \le 10$
H_A:	$\mu > 10$
Critical Values	2.132
Test Statistic	0.393
p value	0.3572

Since p is greater than the level of significance, we cannot reject H_0. There is not enough evidence to say that the mean is more than $10 and prove the boss wrong.

b) This is a two-sided test. We simply want to know if $10 is reasonable.
$\overline{X} = 5.146$ and $s = 1.598$

H_0:	$\mu = 10$
H_A:	$\mu \ne 10$
Critical Values	± 2.776
Test Statistic	-6.791
p value	0.0025

Reject H_0. The mean is not $10, the expense is not justified.

c) Without Tokyo:

Aspirin: $\overline{X} = 6.49$ and $s = 3.70$. You would never claim it is more than $10. To test if it is less than $10, reverse the hypotheses. The test statistic is -1.898 and the p value is 0.076. You do not reject H_0 and there is not enough evidence to say that the boss is correct. $10 would be justified (but barely).

Fast Food: $\overline{X} = 4.53$ and $s = 0.925$. The test statistic is -11.832 and the p value is 0.001. You still can't claim $10. (In fact, here, Tokyo was helping you).
The outlier affects both the sample statistic and the standard error.

9.3 The hypotheses are:
H_0: $\mu \ge 180$
H_A: $\mu < 180$

The output from Minitab is:

T-Test of the Mean

Test of mu = 180.0 vs mu < 180.0

Variable	N	Mean	StDev	SE Mean	T	P
1996	8	174.3	128.8	45.5	-0.13	0.45

Since p = 0.45 is larger than any level of significance that we would choose, we cannot reject H_0. There is not enough evidence to say the mean enrollment has decreased.

9.5 The hypotheses are:
H_0: $\mu \le 5$
H_A: $\mu > 5$

The Minitab output is:

T-Test of the Mean

Test of mu = 5.000 vs mu > 5.000

Variable	N	Mean	StDev	SE Mean	T	P
Hours	25	5.132	0.867	0.173	0.76	0.23

Since the p value is 0.23, which is larger than any p value that we would use, we cannot reject H_0. There is not enough evidence to say that the average amount of time spent web surfing is more than 5 hours per week.

Section 9.3 Exercises

9.7   a) $s^2 = 40.4556$
   b) Ho: $\sigma^2 \geq 100$
      Ha: $\sigma^2 < 100$
   c) The sample size is 10, so there are 9 degrees of freedom.
   Test statistic = 3.640. The critical values is $\chi^2_{0.005, 9} = 16.919$.
   Since the test statistic is not outside the critical values, we cannot reject H_0. There is not enough evidence to say that the variance of the new test is less than 100.

9.9   a) $s^2 = 47.04$
   b) Ho: $\sigma^2 = 50$
      Ha: $\sigma^2 \neq 50$
   There are $24 - 1 = 23$ degrees of freedom.
   Test Statistic = 21.64. The critical values are $\chi^2_{0.025, 23} = 38.076$ and $\chi^2_{0.975, 23} = 11.689$.
   Since the test statistic is not outside these values, we cannot reject H_0. There is not enough evidence to say that the variance of the overruns is greater than 50 (dollars)^2.

Section 9.4 Exercises

9.11   a) and b) The sample proportion was 0.80

H_0:	$p \leq 0.75$
H_A:	$p > 0.75$
Critical Values	1.645
Test Statistic	1.633
p value	0.0512

Since the p value is larger than the level of significance, we cannot reject H_0. There is not enough evidence to say that the percentage is greater than 75%.
c) It is a close call, but they should not implement the programs. They might consider a scaled down version to see if it works, but the data indicate that it is not a good idea.

9.13   a) and b) The sample proportion is 0.493

H_0:	$p = 0.60$
H_A:	$p \neq 0.60$
Critical Values	± 1.96
Test Statistic	-3.771
p value	0.0002

Since the p value is less than the level of significance, we reject H_0. The proportion is not 0.60.
c) Based on the data, the university should change its literature.

Chapter 9 Exercises
Learning It

9.15   a), d) and e) The critical value is $t_{0.15, 24} = 1.711$

H_o:	$\mu \leq 15$
H_A:	$\mu > 15$
Critical Values	1.711
Test Statistic	4.5
p value	0.0000

b) Type I: Say it takes longer than 15 seconds when it does not. This would mean increasing the light time when you should not.

c) Type II: Do not say it is longer than 15 seconds when it is. This would mean not increasing light time when you should.

f) Based on the test, we can conclude that the average time to cross is greater than 15 seconds. If the town really wants to make sure that almost all seniors cross safely, they need to set the light at 19.5 plus about 2 or 3 standard deviations, or at about 30 seconds.

9.17   a), d) and e) The sample proportion is 0.633.

H_o:	$\pi \leq 0.60$
H_A:	$\pi > 0.60$
Critical Values	1.645
Test Statistic	3.727
p value	0.0001

Since p is less than the level of significance, we can reject H_o. The proportion is greater than 0.60 (60%). They should go ahead with the new bottle.

b) Type I: Go ahead with the new bottle when the preference is not greater than 60%.

c) Type II: Do not go ahead with the new bottle when the preference is greater than 60%.

9.19   *First, this is real data, and as such, you notice that it is not simple. You need to leave out the 0s since there can't be attendance if there isn't a game. You also have to make an assumption in calculating the standard deviation. You can't use total attendance there, so you could divide the total by the number of games and use that value as many times as there were games. That will make the standard deviation smaller than it should be, but you don't really have a choice. In this case it won't matter because the standard deviations are HUGE.*

a) $\overline{X} = 6901$ and $s = 2744$; n =12 usable data points

b) $\overline{X} = 8463$ and $s = 2877$; n = 13 usable data points

c) **One-Sample T: Home/Game**
Test of mu = 5000 vs mu not = 5000

Variable	N	Mean	StDev	SE Mean
Home/Game	12	6901	2744	792

Variable	95.0% CI		T	P
Home/Game	(5158,	8645)	2.40	0.035

Since the p value is 0.035, we can reject the null hypothesis and conclude that the average attendance is different than 5000. If we used a different estimate of the standard deviation, the test conclusion would have been the same since it would be smaller.

d) One-Sample T: Away/Game

```
Test of mu = 8000 vs mu not = 8000

Variable          N       Mean      StDev    SE Mean
Away/Game        13       8463       2887       801

Variable              95.0% CI              T        P
Away/Game    (   6718,   10207)         0.58    0.574
```

Since the p value is 0.574, we cannot reject the null hypothesis and therefore there is no evidence to say that the average attendance is different from 2000 for away games.

e) It might be interesting to know what time the games were. It would be better to have each individual attendance rather that total for all games against a given team. It might be better to look at the percent of capacity of the arena to give an idea whether that matters. Perhaps all of the games were full.

Thinking About It

9.21   a)–c) The p value for the test was 0.0001. This says that essentially no matter what level of significance you use, you will reject Ho and conclude that the percentage who favor the new bottle is greater than 60%.

d) I would still not say to go ahead with the change if it is really expensive. There are other factors to consider. How many of the consumers were already Coke drinkers? Would they gain new customers? Would old customers stop drinking Coke if they don't switch? These are all important questions that are not answered with the data given. Coca-Cola needs to do some additional data collection before they make any big decisions.

9.23   a) H_o: $\pi \geq 0.50$ (note π is the percentage that oppose the proposal)
        H_A: $\pi < 0.50$

b) Since the statement is about the percentage that *oppose* the proposal, the 18% should be lumped with the percentage that favor the proposal.

c) The test statistic is $\dfrac{0.46 - 0.50}{\sqrt{\dfrac{0.50 * 0.50}{450}}} = -1.697$.

d) and e)

Type I error: Say the opposition is less than 50% when in fact it is not. This means that they would endorse the casinos when they should not.

Type II error: Do not say the opposition is less than 50%, when in fact it is. This means they would oppose the casinos when they should not.

If they are conservative, they would not want to make a Type I error. They would test at 0.01. If they are aggressive about casino gambling, they would not want to make a Type II error, so they would try to test at 0.10.

f) and g)

H_o:	$\pi \geq 0.50$
H_A:	$\pi < 0.50$
Critical Values	2.33
Test Statistic	-1.697
p value	0.0448

At the 0.01 level of significance, they would not reject H_o. They would not be able to say that opposition was less than 50%, so they would not endorse casino gambling.

h) If we set the level of significance to 0.10, we would reject H_o and conclude that the proportion who oppose it is in fact less than 50%. They would endorse casino gambling.

i) This data really puts you on the fence. I would recommend another poll before I would do anything. A larger sample size might be good.

Chapter 10 Comparing Two Populations - Solutions

Section 10.4 Exercises

10.1   a)–c)

H_o:	$\mu_1 = \mu_2$
H_A:	$\mu_1 \neq \mu_2$
Critical Values	1.96
Test Statistic	4.326
p value	0.0000

d) Since p is less than 0.05 we reject H_o and conclude that the average is different for men and women.

e) In this case, p is zero, so the choice of alpha is not an issue.

10.3   a)–c) Note: Be careful about which population is which (2 is At Home).

H_o:	$\mu_1 \leq \mu_2$
H_A:	$\mu_1 > \mu_2$
Critical Values	1.645
Test Statistic	1.837
p value	0.0331

At the 0.05 level of significance, the data indicate that the average number of binge drinking episodes per semester is less for those students who live at home.

10.5   a)–b)

H_o:	$\mu_1 \leq \mu_2$
H_A:	$\mu_1 > \mu_2$
Critical Values	1.645
Test Statistic	6.595
p value	0.0000

c) The data indicate that the houses listed with Agency A are on the market longer on the average than those listed with Agency B. The difference in the two averages is almost 4 weeks, or a month, so I would use Agency B.

Section 10.5 Exercises

10.7   a)–c) Since we cannot assume equal variances, we need to calculate the degrees of freedom. We truncate the answer rather than round, to be a little more conservative.

$$\frac{\left(\dfrac{0.9^2}{20} + \dfrac{2.3^2}{20}\right)^2}{\dfrac{\left(0.9^2/20\right)^2}{20-1} + \dfrac{\left(2.3^2/20\right)^2}{20-1}} = 24.69 \approx 24$$

H_o:	$\mu_1 \leq \mu_2$
H_A:	$\mu_1 > \mu_2$
Critical Values	1.711
Test Statistic	0.362
p value	0.360

Since p is greater than the level of significance, we cannot reject H_o. There is not enough evidence to say that the USPS is better (faster).

10.9   a) Again, we must calculate degrees of freedom. We truncate the answer rather than round, to be a little more conservative.

$$df = \frac{\left(\dfrac{3.4^2}{10} + \dfrac{1.6^2}{10}\right)^2}{\dfrac{\left(3.4^2/10\right)^2}{10-1} + \dfrac{\left(1.6^2/10\right)^2}{10-1}} = 12.80 \approx 12$$

H_o:	$\mu_1 \le \mu_2$
H_A:	$\mu_1 > \mu_2$
Critical Values	-1.7709
Test Statistic	1.262
p value	0.1147

b) At the 0.05 level of significance, there is no evidence that the time on hold with the new company is better. They should not switch.

Section 10.7 Exercises

10.11   a)

Top Ten Business Software	Computability Price ($)	PC Connection Price ($)	Difference C - PC
Norton Anti-Virus 2000 v 6.0	29	32	-3
Microsoft W98 Second Edition Upgrade	95	90	5
Norton System Works 2000 v 3.0	59	60	-1
VirusScan 5.0	29	24	5
QuickBooks 2000 Pro	200	200	0
Norton Internet Security 2000	58	48	10
QuickBooks 2000	120	120	0
Microsoft W98 Second Edition	180	179	1
Microsoft Office 2000 Upgrade	220	230	-10
VirusScan 5.0 Deluxe	38	33	5

Just looking at the differences, about half are positive and half are negative, so it does not seem that there is really a difference.
b)

	Computability	PC Connection	Difference
Average	102.8	101.6	1.2
Standard Deviation	73.42237	76.60026	5.49343042

c) H_o: $\mu_D = 0$
   H_A: $\mu_D \ne 0$

d) The test statistic is $\dfrac{1.2 - 0}{\dfrac{5.5}{\sqrt{10}}} = 0.6900$. The critical value with 9 degrees of freedom is +/- 2.262, so we cannot

reject Ho. There is no evidence of a difference in average price.
e) The answer does agree with the previous method.

10.13 a)–c) Based on the data, it does appear as if the print sales might be a little higher.

Print Sales	Radio Sales	Difference
28.3	22.1	6.2
24.6	19.1	5.5
23.1	20.3	2.8
21.0	24.4	-3.4
25.7	22.4	3.3
22.5	19.2	3.3
32.0	22.8	9.2
23.5	20.3	3.2
24.3	25.5	-1.2
25.2	22.6	2.6
23.3	24.9	-1.6
25.3	29.7	-4.4
22.2	22.2	0
23.4	28.5	-5.1
23.9	28.2	-4.3
25.7	21.6	4.1
	Average	1.26
	Std. Dev.	4.25

d)

H_o:	$\mu_D = 0$
H_A:	$\mu_D \neq 0$
Critical Values	± 2.750
Test Statistic	1.68
p value	0.1028

e) Since p is greater than the level of significance, we cannot reject H_o. There is not enough evidence to say the two campaigns had different sales.

10.15 a) and b) Looking at the differences, it seems that men use more e-mails than women, because the differences are all positive.

Men	Women	Difference
82	48	34
77	61	16
78	56	22
83	59	24
82	58	24
78	56	22
81	60	21
74	64	10
86	59	27
76	63	13
	Average	21.3
	Std.Dev.	6.9

c)

H_o:	$\mu_D \leq 0$
H_A:	$\mu_D > 0$
Critical Values	1.8331
Test Statistic	9.699
p value	0.0000

d) Since p is less than 0.05, we can reject H_o and conclude that men use more business e-mails than women on the average.

Section 10.8 Exercises

10.17   a) Smokers:: $\hat{p}_1 = 0.0016$ Non-smokers: $\hat{p}_2 = 0.0015$

b) H_o: $p_1 \leq p_2$
   H_A: $p_1 > p_2$

c) Since the p value of the test is 0.2426, we cannot reject H_o. There is no evidence to say that there is a difference in the proportion of deaths from breast cancer.

H_o:	$p_1 \leq p_2$
H_A:	$p_1 > p_2$
Critical Values	1.645
Pooled p	0.0015
Test Statistic	0.6981
p value	0.2426

10.19   a) 1993: $\hat{p}_1 = 0.87$; 1994: $\hat{p}_2 = 0.90$

b)

H_o:	$p_1 \geq p_2$
H_A:	$p_1 < p_2$
Critical Values	-1.645
Pooled p	0.885
Test Statistic	-0.6649
p value	0.2530

c) Since p is greater than 0.05 we cannot reject Ho. There is not enough evidence for Amtrak to say they improved.

Section 10.9 Exercises

10.21   a) Numerator and denominator degrees of freedom are both 14.

H_o:	$\sigma^2_1 = \sigma^2_2$
H_A:	$\sigma^2_1 \neq \sigma^2_2$
F0.95, df1, df2 (lower)	0.403
F0.05, df1, df2 (upper)	2.484
Test Statistic	0.9140

b) Since the test statistic is not outside the critical values, we cannot reject Ho. The assumption of equal variances is ok.

c) Since one of the things you want to be playing golf is consistent, the variability should matter. If one brand were significantly more variable than the other, you might want to construct confidence intervals to see how they compare.

10.23 a) and b)

H_o:	$\sigma^2_1 \leq \sigma^2_2$
H_A:	$\sigma^2_1 > \sigma^2_2$
F0.95, df1, df2 (lower)	none
F0.05, df1, df2 (upper)	1.7045
Test Statistic	3.15

c) Since the test statistic is outside the critical value, we reject H_o and conclude that the variance of Machine A is greater than the variance of Machine B. The quality manager is correct.

Chapter 10 Exercises
Learning It

10.25 a)

$$s^2_p = \frac{(11)(2200^2)+(11)(2100^2)}{12+12-2} = 4624994.34$$

H_o:	$\mu_1 \leq \mu_2$
H_A:	$\mu_1 > \mu_2$
Critical Values	2.508
$s^2 p$	4624994 (sp = 2150.58)
Test Statistic	49.082
p value	0.0000

b) Since the test statistic is outside the critical value, there is evidence that the mean for their city is greater than for the other city. They should lobby for tax incentives.

10.27 a) and b) Looking at the differences, it appears that the time makes a difference since all but one of the differences are positive.

Observation	2.5 sec.	15 sec.	Differences
1	66	78	-12
2	132	115	17
3	120	93	27
4	187	48	139
5	190	77	113
6	17	3	14
7	33	12	21
8	92	12	80
9	1000	146	854
		Average	139.2222222
		Std. Dev.	272.7543665

c) and d)

H$_o$:	$\mu_D \leq 0$
H$_A$:	$\mu_D > 0$
Critical Values	-1.860
Test Statistic	1.53
p value	0.082

Since the test statistic is not outside the critical value, we cannot reject H$_o$. There is not enough evidence to say that the mean number of bacteria has decreased.

e) It is possible that the number of bacteria left on the hands after the first washing affected the results of the second washing.

Thinking About It

10.29   a) To decide what level of significance to use, we should look at the hypotheses and Type I and Type II errors.

H$_o$: $\mu_1 \geq \mu_2$
H$_A$: $\mu_1 > \mu_2$

Type I: Say surgical is better when it is not.
Type II: Don't say surgical is better when it is.

The Type II error seems slightly worse since if the surgical truly is better, people could get well more quickly. Thus, 0.10 would be appropriate.

b) To do the test, we need to know whether or not we can assume the variances are equal, since the sample sizes are small. The summary statistics are shown below:

Descriptive Statistics: LOS (days) by Service

```
Variable    Service           N       Mean     Median    TrMean       StDev
LOS (day    Medical          19      6.421      6.000     6.176       3.820
            Surgical         14      6.929      6.000     6.750       2.921


Variable    Service     SE Mean    Minimum    Maximum         Q1          Q3
LOS (day    Medical       0.876      2.000     15.000      3.000       8.000
            Surgical      0.781      3.000     13.000      4.750       9.250
```

Minitab output for the results of an F test for normal distributions is:

```
F-Test (normal distribution)


Test Statistic: 1.710
P-Value        : 0.164
```

The test statistic for the F test is: $F = \dfrac{3.820^2}{2.921^2} = 1.710$. The critical values (with 18 degrees of freedom in the numerator and 13 in the denominator) are 0.432. and 2.484. Since the test statistic is outside the critical values, we cannot reject H$_0$ so we will assume the variances are equal. This is confirmed by the p value of the test from Minitab.

Now, using Mintab again for a test for independent sample means with variances not assumed equal: (Mintab put Medical as the first population so the alternative hypothesis is less than)

Two-Sample T-Test and CI: LOS (days), Service
```
Two-sample T for LOS (days)


Service    N     Mean    StDev   SE Mean
Medical   19     6.42    3.82    0.88
Surgical  14     6.93    2.92    0.78


Difference = mu (Medical ) - mu (Surgical)
Estimate for difference:  -0.51
90% upper bound for difference: 1.03
T-Test of difference = 0 (vs <): T-Value = -0.43  P-Value = 0.334  DF = 30
```

Since the p value is 0.334 we cannot reject the null hypothesis. There is no evidence that the medical stay is shorter.

c) The p value of the test was 0.334 so the level of significance made no difference in the decision.

d) From the boxplots we see that both of the samples are a bit skewed, the Medical more so, but the normality assumption is not outrageous.

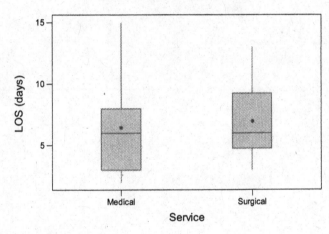

Boxplots of LOS (day by Service)

(means are indicated by solid circles)

10.31 a) H_o: $\mu_1 \geq \mu_2$

H_A: $\mu_1 < \mu_2$

b) Since the sample sizes are both greater than 30, you can use the Z test for two population means.

c)

Ho:	$\mu 1 \geq \mu 2$
HA:	$\mu 1 < \mu 2$
Critical Values	1.645
Test Statistic	-3.944
p value	0.0000

d) Since the p value is less than the level of significance, we reject Ho and conclude that the mean stay at Hospital 1 is shorter.

10.33 a) The p value for the test was 0.0017. The level of significance was not a factor in the decision.

b) If you wanted to test if they were different, the critical value of the test would have changed to ± 1.645. The decision (reject H_o) would not have changed. The conclusion would be that the averages were different, not that the men spent more.

10.35 a) Without pooling the variances, the test statistic is =1.82 and the p value is 0.11. We cannot conclude that the two departments are different, on the average.

b) If we pool the variances, the test statistic changes to - 2.00, but more importantly, the number of degrees of freedom goes up to 16. The p value of the test is 0.063. At the 0.10 level of significance, we would reject Ho and conclude that the departments were different.

c) The change in the number of degrees of freedom can affect the critical value of the test and the results. You need to know which test is appropriate.

Chapter 11 Regression Analysis - Solutions

Section 11.2 Exercises

11.1   a) A linear relationship might be appropriate.

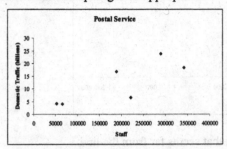

b) The regression equation is $\hat{y} = 0.96 + 0.00006x$

c) Netherlands: $\hat{y} = 0.96 + (0.00006)(53560) = 4.17$ billion

d) France: $\hat{y} = 0.96 + (0.00006)(289156) = 18.30$ billion

$y - \hat{y} = 23.87 - 18.30 = 5.57$

Germany: $\hat{y} = 0.96 + (0.00006)(342413) = 21.50$ billion

$y - \hat{y} = 18.32 - 21.50 = -3.18$

11.3   a)

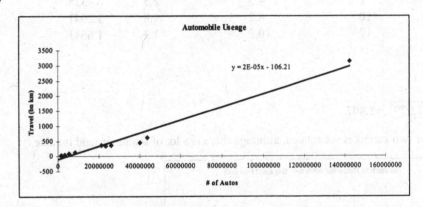

b) $\hat{y} = -106.21 + 0.000022x$. The slope says that for each addition automobile, the miles traveled increases by 0.000022 billion km. (but realize that this is 22,000 km)

c) The intercept does not make sense. If there are no cars, there should be no travel. It should not be negative.

d) See graph above. The fit is really not very good. The point with the high x value (US) seems to be influencing the slope of the line.

e) Sweden: $\hat{y} = -106.21 + (0.000022)(3321000) = -33.15$ (This is REALLY bad)

Japan: $\hat{y} = -106.21 + (0.000022)(40245600) = 779.19$. This is almost double the actual value.

11.5   a–c The regression equation is $\hat{y} = 0.7077 + 0.00001x$

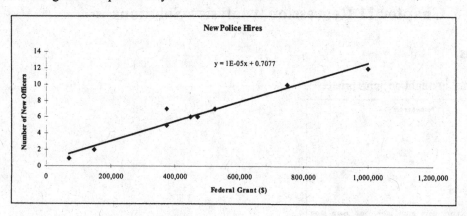

d) The line fits the data pretty well. There is only one point that seems far from the line.

e)

Grant ($)	# Officers	Predicted	Residuals (e)	e^2
150,000	2	2.2	-0.2	0.0441
375,000	5	4.5	0.5	0.2916
471,125	6	5.4	0.6	0.3350
70,967	1	1.4	-0.4	0.1761
450,000	6	5.2	0.8	0.6241
525,000	7	6.0	1.0	1.0816
375,370	7	4.5	2.5	6.4328
750,000	10	8.2	1.8	3.2041
1,000,000	12	10.7	1.3	1.6641

f)

$$s_{y|x} = \sqrt{\frac{13.8537}{7}} = \sqrt{1.9791} = 1.407$$

11.7   a) and b) It appears that the two variables are related, although there is a lot of scatter around the line.

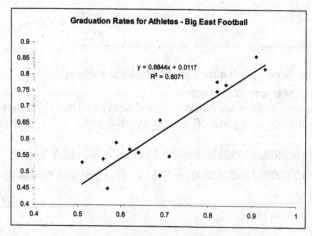

c) Virginia Tech: $\hat{y} = 0.0117 + (0.8844)(0.69) = 0.622$

   Providence: $\hat{y} = 0.0117 + (0.8844)(0.82) = 0.737$

d) It does seem a little better. There is a little less scatter around the line. The new line is $\hat{y} = 0.8743x + 0.0291$.

Section 11.3 Exercises

11.9   a) $\dfrac{0.937}{\sqrt{3118.04 - \dfrac{396.2^2}{52}}} = 0.0940$

b) H_o: $\beta_1 = 0$
H_A: $\beta_1 \neq 0$

$t = \dfrac{0.442 - 0}{3.580} = 0.123$ $t = \dfrac{0.808 - 0}{.0940} = 8.596$

c) At the 0.10 level of significance, the critical value is $t_{0.05,\ 50} = -1.676$ (which is VERY close to the Z value). Since 8.596 is very much outside the critical value, we reject H_o. The slope is not zero.

11.11   a) The slope, b_1 is 0.000022, $s_{b_1} = 0.0000012$, $t = 18.07$ and the p value of the test is 0 (for all practical purposes).

b) Since 0 is less than 0.01, we can reject H_o and conclude that the relationship between number of vehicles and distance traveled is significant.

c) The critical value is $t_{0.01,\ 10} = 2.764$. The t statistic is very much outside the critical value, which verifies the result of part b.

11.13   Looking at the p value for the slope coefficient we see that it is 0 for all practical purposes. The relationship is significant.

Section 11.4 Exercises

11.15   From Exercise 11.5 the regression equation is $\hat{y} = 0.7077 + 0.00001x$.

For $x = 350,000$, $\hat{y} = 4.21$. $t_{0.025,\ 7} = 2.365$.
The error of the estimate is:

$$(2.365)(0.8033)\sqrt{\dfrac{1}{9} + \dfrac{(350,000 - 463,051)^2}{2,571,674,717,614 - \dfrac{(4,167,462)^2}{9}}} = (2.365)(0.8033)\sqrt{0.131020\ldots} = 0.688$$

so the confidence interval is 4.21 ± 0.688.

The prediction interval is almost the same, with one change in the square root part:

$$(2.365)(0.8033)\sqrt{1 + \dfrac{1}{9} + \dfrac{(350,000 - 463,051)^2}{2,571,674,717,614 - \dfrac{(4,167,462)^2}{9}}} = (2.365)(0.8033)\sqrt{1.131020\ldots} = 2.019$$

so the prediction limits are 4.21 ± 2.019.

11.17   a) and c)

X value	Fit	SE Fit	95.0% CI	95.0% PI s
1000	23.010	0.734	(21.373, 24.646)	(17.548, 28.472)
1500	33.768	1.085	(31.350, 36.187)	(28.023, 39.513)

b) The interval for 1500 is wider because the x value is further from the mean, $\overline{X}$, which is 742.

d) It depends on what you are trying to do with the estimate. If you were using the estimates to plan on a macro level (say for the United States) then the confidence interval is probably best. If you are trying to use the estimate for planning on a micro level (say, for a particular state) like predicting revenue from sales tax, the prediction interval is better.

Section 11.5 Exercises

11.19   a)

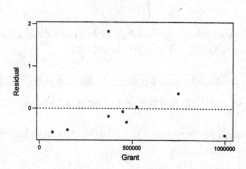

b) There is no pattern to indicate that the relationship is not linear.

c) The points are fairly equally spread around zero with the exception of one observation. There is no obvious increase or decrease in variability as X changes, so equal variances is probably a reasonable assumption.

d) Because of the small number of observations, a boxplot or dotplot would be best. With the exception of the obvious outlier, the data appear to be symmetric.

e) From this plot, it appears that the normality assumption might not be reasonable.

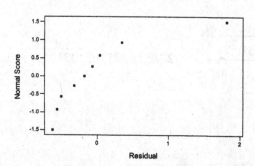

f) The linear model is probably reasonable. There is one observation (the seventh) which seems to be causing some problems. It might be good to drop that observation and see how things change.

11.21   a)

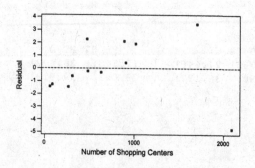

b) and c) There seem to be two problems. One is that the residuals are not randomly distributed around zero, but instead trend upward, indicating that a linear relationship might not be appropriate. Also, it appears that there might be an increase in variability as X increases, although that last point might be an outlier.

d) and e) The data appear to be slightly skewed, or have some other problem.

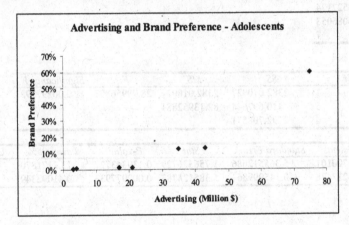

f) No, the linear model is definitely not appropriate here.

Chapter 11 Exercises
Learning It

11.23   a) The independent variable is Advertising and the dependent variable is Preference.

b) There might be a linear relationship between the two variables, but it is not linear.

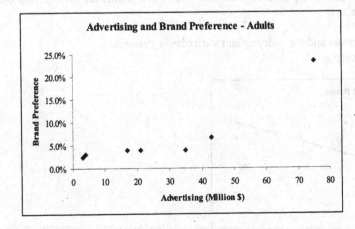

c) It looks like the adolescent brand preference is more strongly related to advertising.

d) The regression equation is $\hat{y} = -9.42 + 0.79x$

e) The intercept says that even if a brand does not advertise, the brand preference will be - 9%. It is not unreasonable that this would not be zero, but it is clear that a negative intercept does not make sense. The slope says that for every 1 million dollars spent, the brand gains 0.79% in market share.

f) The predicted values not very close to the actual values.

Brand	Advertising	Adolescent	Predicted
Marlboro	75	60%	49.5%
Camel	43	13%	24.4%
Newport	35	13%	18.1%
Kool	21	1%	7.1%
Winston	17	1%	4.0%
Benson & Hedges	4	1%	-6.3%
Salem	3	0%	-7.1%

g) The Excel output is:

SUMMARY OUTPUT

Regression Statistics	
Multiple R	0.923547136
R Square	0.852939312
Adjusted R Square	0.823527174
Standard Error	9.063086053
Observations	7

ANOVA

	df	SS	MS	F	Significance F
Regression	1	2382.010927	2382.010927	28.99956893	0.002977933
Residual	5	410.697644	82.13952881		
Total	6	2792.708571			

	Coefficients	Standard Error	t Stat	P-value	Lower 95%	Upper 95%
Intercept	-9.424720101	5.365512686	-1.756536729	0.139343727	-23.21718701	4.36774681
Advertising	0.786227478	0.145999863	5.385124783	0.002977933	0.410923497	1.16153146

From the output, we test:
H_o: $\beta_1 = 0$
H_A: $\beta_1 \neq 0$
We see that the test statistic is 5.385 and the p value is 0.0029. There is a significant relationship between advertising and brand preference.

11.25   a) The dependent variable is deposits and the independent variables is customers.
b and c) The relationship is not very good.

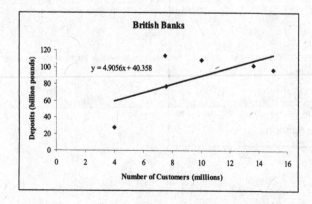

d) The line does not even come close.
e) and f) The standard error is 27.71. The t statistic is 1.633 and the p value is 0.178. The model is not significant.

Thinking About It

11.27 a) The regression equation is $\hat{y}$ = -4.9302 + 0.0303x

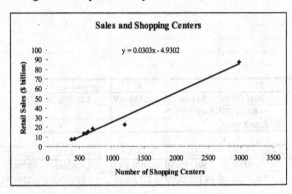

SUMMARY OUTPUT

Regression Statistics	
Multiple R	0.632499359
R Square	0.400055439
Adjusted R Square	0.250069299
Standard Error	27.70702187
Observations	6

ANOVA

	df	SS	MS	F	Significance F
Regression	1	2047.617089	2047.617089	2.667282713	0.177768364
Residual	4	3070.716245	767.6790612		
Total	5	5118.333333			

	Coefficients	Standard Error	t Stat	P-value	Lower 95%	Upper 95%
Intercept	40.35811412	31.02123058	1.30098366	0.263146082	-45.77080809	126.4870363
Customers (million)	4.905568723	3.003688135	1.633181776	0.177768364	-3.434023772	13.24516122

b) The regression equation for the North Central States was: $\hat{y}$ = 0.0215x + 1.4926. The two equations are similar in that the values of the slope and intercept are both the same order of magnitudes. The South Central States have a negative intercept and a slightly higher slope.
c) No, they would probably not be the same. Much would depend on the economy of the regions.
d) There should be similarities in order of magnitude of the slope and intercept terms. The slopes should all be positive. The slopes and intercepts should differ in exact value.

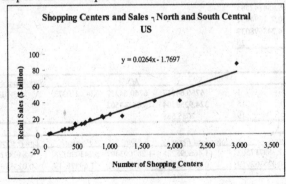

e) and f) From the plot, it looks like more points are far away from the line.
The regression output for all three models are shown below:
North Central:

SUMMARY OUTPUT

Regression Statistics	
Multiple R	0.986695521
R Square	0.973568052
Adjusted R Square	0.970924857
Standard Error	2.338760078
Observations	12

ANOVA

	df	SS	MS	F	Significance F
Regression	1	2014.69118	2014.69118	368.3300409	3.21059E-09
Residual	10	54.69798702	5.469798702		
Total	11	2069.389167			

	Coefficients	Standard Error	t Stat	P-value	Lower 95%	Upper 95%
Intercept	1.492611658	1.0713872	1.393158009	0.193763713	-0.894588201	3.879811518
Number of Shopping Centers	0.021517145	0.001121156	19.19192645	3.21059E-09	0.019019053	0.024015237

South Central:

SUMMARY OUTPUT

Regression Statistics	
Multiple R	0.991158672
R Square	0.982395513
Adjusted R Square	0.979461431
Standard Error	3.770753692
Observations	8

ANOVA

	df	SS	MS	F	Significance F
Regression	1	4760.69725	4760.69725	334.8221911	1.71636E-06
Residual	6	85.31150043	14.21858341		
Total	7	4846.00875			

	Coefficients	Standard Error	t Stat	P-value	Lower 95%	Upper 95%
Intercept	-4.930177714	2.043655753	-2.412430619	0.052400189	-9.930826852	0.070471425
Number of Shopping Centers	0.030272553	0.001654405	18.2981472	1.71636E-06	0.026224366	0.03432074

Combined:

SUMMARY OUTPUT

Regression Statistics	
Multiple R	0.976802224
R Square	0.954142585
Adjusted R Square	0.951594951
Standard Error	4.248698039
Observations	20

ANOVA

	df	SS	MS	F	Significance F
Regression	1	6760.64217	6760.64217	374.5210373	1.6987E-13
Residual	18	324.9258304	18.05143502		
Total	19	7085.568			

	Coefficients	Standard Error	t Stat	P-value	Lower 95%	Upper 95%
Intercept	-1.769717503	1.466098225	-1.207093407	0.243020128	-4.849877961	1.310442956
Number of Shopping Centers	0.026362959	0.001362248	19.35254602	1.6987E-13	0.023500981	0.029224937

All three models are significant, but the standard error for the combined model is the largest. Despite the larger value of n (20 vs 12 and 8) the prediction and confidence interval estimates might be less precise. Unless there was a reason to combine them (similar economies, planning) the individual models might be better.

11.29   a) and b)

New Obs	Fit	SE Fit	95.0% CI	95.0% PI
500000	362.5	20.6	(321.1, 403.9)	(81.0, 644.1)
1000000	743.8	19.5	(704.7, 782.9)	(462.6, 1025.1)
2000000	1506.4	26.7	(1452.8, 1560.1)	(1222.8, 1790.1)

c) Look at the prediction interval for 500,000 dropouts. The smallest number of people below the poverty level would be estimated at 81.0 thousand (81,000) and the largest estimate would be 644 thousand (644,000). That is off by almost half a million people which is just not useful for planning purposes.

d) Since there is almost a 1 to 1 correspondence (according to the model), a reduction of 10% in the number of dropouts should correspond to a reduction of 10% in the number below the poverty level. However, there is no evidence of cause and effect. That is, there are other factors which might need to be considered before such programs were instituted.

11.31   a) The regression model is $\hat{y} = 51.96 + 0.024x$

b) The test statistic is 1.775 and the p value is 0.150, so at the 0.05 level of significance, the relationship between number of branches and deposits is not significant.

c) R for this model is 44% and the correlation coefficient is 0.6638.

d) Looking at R^2 for each model, the value for the model using number of customers is 40% (correlation coefficient is 0.6325) and for the model using number of branches it is 44% (correlation coefficient is 0.0.6638). The models are not really that different from each other. They are equally bad.

11.33   a) and b) The residuals are not randomly distributed around 0. In fact, only one point is below 0, suggesting a nonlinear model.

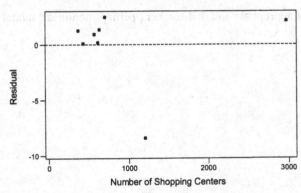

Residuals Versus Number of Shopping Centers

c) and d) The graphs are identical, because X and Y are related.

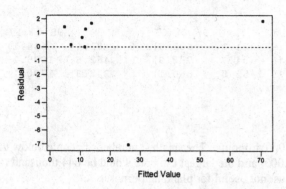

e) The normal probability plot indicates a problem with the normality assumption.

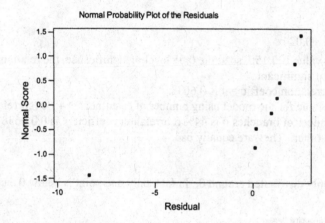

f) It would seem that the linear model is not appropriate and that another (perhaps nonlinear) model should be considered.

Chapter 12 Multiple Regression Models - Solutions

Section 12.2 Exercises

12.1 a) There are 3 independent variables in the model: Licensed Drivers (X_1), Registered Vehicles (X_2) and Vehicle Miles (X_3).

b) The regression model is $\hat{y} = 51.75 + 0.06x_1 - 0.21x_2 + 0.029x_3$.

c) The coefficient of X_1 says that the number of fatalities increases by 0.06 for every increase of 1 in licensed drivers. The number of fatalities decreases by 0.21 for every increase in registered vehicles and increases by 0.029 for every increase in vehicle miles.

d)

State	Fatalities	Population	Licensed Drivers (X_1) (thousands)	Registered Vehicles (X_2) (thousands)	Vehicle Miles (X_3)(millions)	Predicted Fatalities
AL	1,083	4,219	3,043	3,422	48,956	935
AK	85	606	443	508	4,150	92
AZ	9,03	4,075	2,654	2,980	38,774	710
AR	610	2,453	1,770	1,560	24,948	554
CA	4,226	31,431	20,359	23,518	271,943	4221
CO	585	3,656	2,620	3,144	33,705	526

e) If you look at the plot, you see that the points fall very close to the line $\hat{y} = y$. The predictions seem to be good.

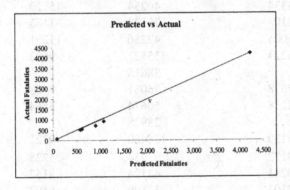

12.3 a) The dependent variable is Serious Crimes. The independent variables are number of police officers in the city, the percent of arrested males who tested positive for drugs, the number of families in the city below the poverty level, the number of people between 16 and 19 who were not enrolled in school and were not high school graduates and the number of single female parent households.

b) The selection of variables is reasonable. There are probably many other factors that might affect the number of serious crimes such as mean or median income level, level of education, house prices, etc.

c) The output from Minitab is shown below.

Regression Analysis

```
The regression equation is
Serious Crimes = 3211 + 12.7 Police Officers + 5.69 Non Enrolled/Not Grad
           + 2.71 Families < Poverty Level
           - 1.98 Single Female Parent Households + 414 Male Arrests Drugs (%)

20 cases used 3 cases contain missing values

Predictor      Coef      StDev        T        P
Constant       3211      36610     0.09    0.931
Police O      12.700      6.732     1.89    0.080
Non Enro       5.686      1.593     3.57    0.003
Families      2.7116     0.8912     3.04    0.009
Single F     -1.9805     0.6679    -2.97    0.010
Male Arr      414.0      612.0      0.68    0.510
```

```
S = 18659       R-Sq = 95.3%     R-Sq(adj) = 93.6%

Analysis of Variance

Source          DF          SS           MS          F         P
Regression       5  98029534093  19605906819      56.32     0.000
Residual Error  14   4874055532    348146824
Total           19  1.02904E+11

Source      DF      Seq SS
Police O     1  67197173082
Non Enro     1  27087595886
Families     1    418194898
Single F     1   3167264519
Male Arr     1    159305708
```

d) The predictions are best for Indianapolis, IN and worst for Dallas, TX.

CITY	Serious Crimes	Predicted	Residual
Atlanta, GA	76398	56072	20306
Birmingham, AL	33895	43474	-9580
Chicago, IL	*	269027	
Cleveland, OH	45610	75946	-30336
Dallas, TX	154929	112102	42827
Denver, CO	36558	55303	-18745
Detroit, MI	127080	123401	3679
Fort Lauderdale, FL	24334	40254	-15920
Houston, TX	180308	179173	1135
Indianapolis, IN	36005	47256	-11251
Los Angeles, CA	346224	355827	-9603
Manhattan, NY	*	38618	
Miami, FL	67678	76081	-8403
New Orleans, LA	54238	50674	3564
Omaha, NE	*	28928	
Philadelphia, PA	109139	118460	-9321
Phoenix, AZ	99172	93934	5238
Portland, OR	50281	42124	8157
St. Louis, MO	64103	57800	6303
San Antonio, TX	117486	109874	7612
San Diego, CA	96781	77812	18969
San Jose, CA	42836	49842	-7006
Washington, DC	64393	62044	2349

e) The predictions are in the table above. The prediction for Chicago is the biggest number in the data. The prediction for Manhattan, NY seems VERY low.

f) The line seems to do a pretty good job of predicting, although it didn't seem that way looking at the data.

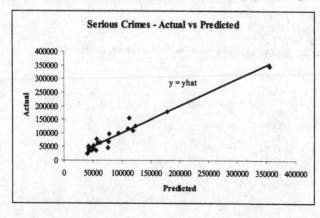

Section 12.3 Exercises

12.5   a) H_o: $\beta_1 = \beta_2 = \beta_3 = 0$
      H_A: at least one β is not 0
   b) From the output, the F statistic is 437.149 and the p value is 0.0000. The model is significant.
   c) The value of $R^2 = 0.965$ (96.5%)
   d) The model explains 96.5% of the variation in Y. It is a good model.
   e) and f) H_o: $\beta_i = 0$
          H_A: $\beta_i \neq 0$ for i = 1, 2, 3

Variable	Test statistic	p value	Significant?
X_1 (Licensed Drivers)	1.289	0.0957	no
X_2 (Registered Vehicles)	-3.785	0.0004	yes
X_3 (Vehicle Miles)	8.326	0.0000	yes

   g) The coefficient of X_2 is significantly different from zero.
   h) The slope of X_1 is not significantly different from zero, but X_2 and X_3 are.
   i) Use the model with X_2 and X_3 in it.

12.7   a) H_o: $\beta_1 = \beta_2 = \beta_3 = \beta_4 = \beta_5 = 0$
      H_A: at least one β is not 0
   b) **Regression Analysis**

```
The regression equation is
Serious Crimes = 3211 + 12.7 Police Officers + 5.69 Non Enrolled/Not Grad
          + 2.71 Families < Poverty Level
          - 1.98 Single Female Parent Households + 414 Male Arrests Drugs (%)

20 cases used 3 cases contain missing values

Predictor      Coef       StDev         T       P
Constant       3211       36610      0.09    0.931
Police O      12.700       6.732      1.89    0.080
Non Enro       5.686       1.593      3.57    0.003
Families      2.7116      0.8912      3.04    0.009
Single F     -1.9805      0.6679     -2.97    0.010
Male Arr       414.0       612.0      0.68    0.510

S = 18659      R-Sq = 95.3%     R-Sq(adj) = 93.6%

Analysis of Variance

Source          DF          SS           MS         F       P
Regression       5 98029534093  19605906819     56.32   0.000
Residual Error  14  4874055532    348146824
Total           19 1.02904E+11
```

Since the F statistic is 56.32 and the p value is 0.000, the model is significant. At least one of the variables is significant.
   c) The R^2 value is 95.3%.
   d) H_o: $\beta_i = 0$
        H_A: $\beta_i \neq 0$ for i = 1, 2, 3, 4, 5

e)

Variable	Test statistic	p value	Significant?
Police Officers	1.89	0.080	no
Not Enrolled/Non Grad	3.57	0.003	yes
Families < Poverty	3.04	0.009	yes
Single Female Parent	-2.97	0.010	maybe yes, maybe no
Male Arrests (Drugs)	0.68	0.510	no

At least two of the variables are significant. The number of Single Female Parent Homes is a judgement call. I would recommend trying both models and if the R^2 value is much higher for the three variable model, I would include all three. Otherwise I would pick the model with the other two significant variables.

12.9   a) The new model is $\hat{y} = 46.04 - 0.163x_1 + 0.030x_2$

b) and c) When the number of licensed drivers is dropped, the R^2 value decreases from 96.5% to 71.3%. The model is still significant. The coefficients of the variables and the intercept change slightly but they keep the same sign. If you don't keep the other two variables, it might be good to look for some others.

d) The two variable model is still a good model. The coefficients of the two variables that were dropped were very small and did not really change $\hat{y}$.

e) I would try a few more variables.

Chapter 12 Exercises
Learning It

12.11   a) The regression model is $\hat{y} = 25.02 - 1.19x_1 - 0.48x_2$; where X_1 = Debt and X_2 = Employees. The negative coefficients mean that equity decreases with debt (which makes sense) and the number of employees (not sure why that would be true).
b) and c)

Year	Equity	Predicted Equity	Residuals
1986	7.55	7.21	0.34
1987	7.08	7.35	-0.27
1988	8.69	9.13	-0.44
1989	8.74	9.82	-1.08
1990	10.51	9.65	0.86
1991	10.61	9.39	1.22
1992	10.37	10.25	0.12
1993	10.28	11.92	-1.64
1994	11.29	12.52	-1.23
1995	12.16	13.00	-0.84
1996	16.15	13.21	2.94

The largest residual is in 1996 and the smallest is in 1992.
d) R^2 for the model is 71.9%. This is not too bad, but if you look at the individual coefficients, you see that neither one is significant - most of the model is in the intercept term. This means that the model is not very good.
e) and f) H_o: $\beta_1 = \beta_2 = 0$
         H_A: at least one β is not 0
  Since the F statistic is 10.257 and the p value is 0.006, the model is significant.
g) The R^2 value is 71.9%, which says that the model containing the variables debt and employees can explain 71.9% of the variation in equity.
h) The model using revenue and assets had an R^2 of 85.2%, and the individual variable revenue was significant. It is a better model for both reasons.

12.13 a)

SUMMARY OUTPUT

Regression Statistics	
Multiple R	0.916393955
R Square	0.839777881
Adjusted R Square	0.825212234
Standard Error	5167.075596
Observations	49

ANOVA

	df	SS	MS	F	Significance F
Regression	4	6157214287	1539303572	57.65469063	6.21404E-17
Residual	44	1174741490	26698670.22		
Total	48	7331955776			

	Coefficients	Standard Error	t Stat	P-value	Lower 95%	Upper 95%
Intercept	1518.152586	1088.370917	1.394885294	0.17005261	-675.3147864	3711.619959
Acreage	-1.488594986	1.562806966	-0.952513662	0.346040065	-4.638225355	1.661035382
Day Visitors	-0.237309684	0.079686615	-2.978036968	0.004704979	-0.397907497	-0.076711871
Overnight Visitors	2.845806639	0.844106732	3.371382468	0.001566857	1.144621364	4.546991914
Operating Expenses	0.384472825	0.04422506	8.693551266	4.16561E-11	0.295343077	0.473602574

b) The model is $\hat{y} = 1518.2 - 1.49x_1 - 0.237x_2 + 2.85x_3 + 0.384x_4$. The coefficient for operating expenses is positive, says that the more it costs to operate the park the more the revenue. As number of acres and day visitors goes up, the revenue goes down which doesn't really make sense. The coefficient for overnight visitors is positive, which is encouraging.

c) and d)

State	Total Revenue	Predicted	Residuals	State	Total Revenue	Predicted	Residuals
Alabama	25,724	14535	11,189	Nebraska	11,003	9021	1,982
Alaska	1,957	-130	2,087	Nevada	1,362	3479	-2,117
Arizona	4,114	7072	-2,958	New Hampshire	5,492	3828	1,664
Arkansas	12,805	10724	2,081	New Jersey	6,914	8967	-2,053
California	63,689	69109	-5,420	New Mexico	3,345	14554	-11,209
Colorado	10,708	6293	4,415	New York	42,299	42306	-7
Connecticut	3,571	3864	-293	North Carolina	2,481	5391	-2,910
Delaware	5,013	4813	200	North Dakota	803	2490	-1,687
Florida	21,606	18744	2,862	Ohio	20,997	18166	2,831
Georgia	16,917	21175	-4,258	Oklahoma	16,679	13543	3,136
Idaho	2,801	4446	-1,645	Oregon	12,185	8345	3,840
Illinois	4,282	6993	-2,711	Pennsylvania	9,350	21146	-11,796
Indiana	23,120	17744	5,376	Rhode Island	3,039	4196	-1,157
Iowa	2,700	3677	-977	South Carolina	14,431	10881	3,550
Kansas	3,057	5040	-1,983	South Dakota	5,923	4565	1,358
Kentucky	42,260	26338	15,922	Tennessee	24,538	14951	9,587
Louisiana	3,598	5807	-2,209	Texas	18,822	13927	4,895
Maine	1,789	3139	-1,350	Utah	5,494	9454	-3,960
Maryland	11,163	10770	393	Vermont	4,963	4423	540
Massachusetts	3,310	12661	-9,351	Virginia	3,978	6293	-2,315
Michigan	24,143	23594	549	Washington	9,337	7690	1,647
Minnesota	9,201	9883	-682	West Virginia	15,250	11177	4,073
Mississippi	5,442	7277	-1,835	Wisconsin	10,230	7111	3,119
Missouri	6,100	10966	-4,866	Wyoming	643	5595	-4,952
Montana	1,126	3723	-2,597				

e) The largest residual is California, and the smallest is New York.

f) R^2 is 83.98%. The model does a good job of predicting revenues, even if the coefficients make no sense.

Thinking About It

12.15   a) **Regression Analysis**

The regression equation is
Shareholders Equity ($/share) = 9.5 +0.000817 Assets ($ millions)
          +0.000192 Revenue ($millions)

6 cases used 5 cases contain missing values

Predictor	Coef	StDev	T	P
Constant	9.51	15.39	0.62	0.580
Assets (	0.0008170	0.0007539	1.08	0.358
Revenue	0.00019250	0.00009826	1.96	0.145

S = 1.184   R-Sq = 83.5%   R-Sq(adj) = 72.5%

Analysis of Variance

Source	DF	SS	MS	F	P
Regression	2	21.250	10.625	7.58	0.067
Residual Error	3	4.204	1.401		
Total	5	25.454			

Source	DF	Seq SS
Assets (	1	15.872
Revenue	1	5.379

b) I would expect the signs of the coefficients to be the same, and I would expect the same variables to be important. The values of the coefficients would be different, depending on the magnitude of the data.

c) The signs of the two coefficients are the same. The intercept is different (and you would not expect them to be the same), but the coefficients here are much smaller numbers. The R^2 value is about the same, but the model for Mobil is not significant.

d) **Regression Analysis**

The regression equation is
Shareholders Equity ($/share) = 62.5 - 0.00247 Total Debt ($millions)
          + 0.001 Employees (1000)

6 cases used 5 cases contain missing values

Predictor	Coef	StDev	T	P
Constant	62.537	8.634	7.24	0.005
Total De	-0.002470	0.002643	-0.93	0.419
Employee	0.0014	0.2302	0.01	0.996

S = 1.447   R-Sq = 75.3%   R-Sq(adj) = 58.9%

Analysis of Variance

Source	DF	SS	MS	F	P
Regression	2	19.172	9.586	4.58	0.123
Residual Error	3	6.282	2.094		
Total	5	25.454			

Source	DF	Seq SS
Total De	1	19.172
Employee	1	0.000

e) The coefficient of debt is the same sign (negative), but the one for employees is the opposite. The R^2 value for both models are comparable.

f) The model with assets and revenue is better, but neither one is significant.

12.17 a) The R^2 value is 83.5%. This says that 83.5% of the variation in sales can be accounted for by the model containing these variables.

b) The model should be useful in determining which factors affect sales and in predicting the sales.

c) H_o: $\beta_i = 0$

H_A: $\beta_i \neq 0$ for i = 1, 2, 3, 4, 5

Variable	Test statistic	p value	Significant?
Store Size	3.06	0.006	yes
Windows	0.31	0.756	no
Competitors	-3.08	0.006	yes
Mall Size	-3.22	0.005	yes
Nearest Competitor	1.31	0.204	no

d) If I were to drop one variable, it would be Windows, since it has the smallest t score.

12.19 a) Refer to the output in Exercise 12.12

Variable	Test statistic	p value	Significant?
Advertising	1.559	0.1354	no
Competitors	-3.137	0.0054	yes
Discounts	0.672	0.5095	no
Warranty	2.602	0.0175	yes

b) Two of the variables are significant, number of competitors and warranty. The other two are not.

c) I would drop discounts since it had the smallest t value.

d)

SUMMARY OUTPUT

Regression Statistics	
Multiple R	0.893929099
R Square	0.799109235
Adjusted R Square	0.76897562
Standard Error	300.5692202
Observations	24

ANOVA

	df	SS	MS	F	Significance F
Regression	3	7187290.211	2395763.404	26.51886408	3.58101E-07
Residual	20	1806837.123	90341.85613		
Total	23	8994127.333			

	Coefficients	Standard Error	t Stat	P-value	Lower 95%
Intercept	3997.973577	726.2335721	5.505079538	2.17728E-05	2483.077596
Advertising ($)	1.900322452	1.329021908	1.429865408	0.168192273	-0.871967381
# Competitors	-411.3369915	68.6078719	-5.995478071	7.31596E-06	-554.450438
Warranty (years)	253.7434497	99.53235387	2.549356464	0.019103031	46.12269412

e) The model does not really change very much. The coefficients change slightly, but the basic model is the same. The R^2 value changes from 80.3% to 79.9% which is almost not at all.

f) I would still consider dropping advertising, since it is still not significant.

g)

SUMMARY OUTPUT

Regression Statistics	
Multiple R	0.882367851
R Square	0.778573025
Adjusted R Square	0.757484741
Standard Error	307.9534914
Observations	24

ANOVA

	df	SS	MS	F	Significance F
Regression	2	7002584.923	3501292.462	36.91969667	1.33328E-07
Residual	21	1991542.41	94835.35287		
Total	23	8994127.333			

	Coefficients	Standard Error	t Stat	P-value	Lower 95%	Upper 95%
Intercept	5011.90499	160.6274097	31.20205326	4.44662E-19	4677.861948	5345.948033
# Competitors	-465.7024952	58.5118021	-7.959120698	8.93934E-08	-587.38447	-344.0205204
Warranty (years)	250.7466411	101.9550161	2.459385038	0.022678061	38.71954138	462.7737408

h) I would use the two variable model. The value of R^2 only drops to 77.8%, so the three variable model is not much better. If they want a better model, they will have to find addition different variable(s) to use.

i) This is very similar to Backwards Regression.

12.21 a) Best Subsets Regression

Response is Sales ($

```
                              A N D W
                              d o i a
                              v . s r
                              e   s r
                              r C c r
                      Adj.    t o a n
Vars  R-Sq   R-Sq    C-p    s i u n t
                              m n t y
                              s p t y

1    71.5   70.2    7.6   341.46    X
1    64.6   63.0   14.2   380.19      X
1    34.1   31.1   43.8   519.14  X
1    11.1    7.0   66.1   602.99          X
2    77.9   75.7    3.4   307.95    X     X
2    73.4   70.8    7.8   337.64  X X
2    71.7   69.0    9.4   348.33    X X
2    68.7   65.8   12.3   365.91      X X
2    65.7   62.4   15.2   383.22  X   X
3    79.9   76.9    3.5   300.57  X X   X
3    77.9   74.5    5.4   315.49    X X X
3    73.4   69.4    9.8   345.95  X X X
3    70.2   65.7   12.8   365.97  X   X X
4    80.4   76.2    5.0   304.77  X X X X
```

Number of Variables	R^2	Variables in Model
1	71.5	No. Competitors
2	77.9	No. Competitors, Warranty
3	79.9	No. Competitors, Warranty, Advertising
4	80.4	All

Since the R^2 values for the 3 and 4 variable model are almost identical, the 3 variable is better. Comparing the 3 variable model to the best 2 variable model, we see that the R^2 value decreases from 79.9% to 77.9%, but the adjusted R^2 values are even closer. Neither model has much bias, but I think I would choose the 2 variable model. Adding advertising does not improve the model enough.

b) The two models are the same. I would not necessarily expect them to be identical, but they should contain the same core set of variables.

c) Since they are the same, it does not really matter. I still like the two variable model.

12.23 a) Refer to the output from Exercise 12.16.

Variable	Test statistic	p value	Significant?
DRVRCNT	-0.1401	0.8889	no
VEHCOUNT	3.374	0.001	yes
NUM KIDS	2.574	0.0116	yes
DTVCNT H	-0.831	0.4078	no
DTVMILH	1.045	0.2986	no

At the 0.05 level of significance, number of vehicles and number of children are significant.

b) I would drop number of drivers and the information about the previous 24 hours.

c) No, households with no vehicles are not part of the population of interest and should not be included.

d) The model has not really changed much. The same variables are still significant and the signs of the coefficients still do not all make sense.

SUMMARY OUTPUT

Regression Statistics	
Multiple R	0.429651821
R Square	0.184600687
Adjusted R Square	0.139300725
Standard Error	20493.08485
Observations	96

ANOVA

	df	*SS*	*MS*	*F*	*Significance F*
Regression	5	8556972929	1711394586	4.075073793	0.002217828
Residual	90	37796987384	419966526.5		
Total	95	46353960313			

	Coefficients	*Standard Error*	*t Stat*	*P-value*	*Lower 95%*
Intercept	4467.904808	5733.066887	0.77932194	0.437834563	-6921.824048
DRVRCNT	-283.8270824	3274.419054	-0.086680134	0.931118326	-6789.026817
VEHCOUNT	7746.509388	2642.944327	2.931014971	0.004282251	2495.843408
NUM_KIDS	5353.346337	2097.589318	2.552142257	0.012392089	1186.122396
DTVCNT_H	-519.6829151	617.5907033	-0.841468164	0.40231601	-1746.633654
DTVMILH	39.82971801	41.54939484	0.958611266	0.340322767	-42.71534047

e) I am not comfortable include households with vehicles but no drivers. I am also uncomfortable with households that have vehicles and/or drivers but drove 0 miles in the year. Some of these have miles for the previous 24 hours, which really makes no sense. Since they are not typical of the population of interest, I would drop them too.

f) Even after dropping these values, the R^2 value is not high enough to be good and the coefficients still do not make sense. The two variable model is the best, but it is not good and I am not comfortable with it.

12.25 The model that is currently best has number of vehicles and number of children included. The output from Minitab is shown below:

Regression Analysis

```
The regression equation is
HHVMILES = 4913 + 7773 VEHCOUNT + 4907 NUM_KIDS

Predictor      Coef      StDev        T       P
Constant       4913      5377      0.91    0.363
VEHCOUNT       7773      2253      3.45    0.001
NUM_KIDS       4907      1904      2.58    0.012

S = 19957      R-Sq = 18.7%      R-Sq(adj) = 16.8%

Analysis of Variance

Source          DF          SS           MS        F       P
Regression       2  7944219074   3972109537     9.97   0.000
Residual Error  87 34651080151    398288278
Total           89 42595299225

Source          DF       Seq SS
VEHCOUNT         1   5298114849
```

```
NUM_KIDS     1   2646104225
```

Unusual Observations

Obs	VEHCOUNT	HHVMILES	Fit	StDev Fit	Residual	St Resid
8	2.00	101313	25368	2125	75945	3.83R
27	3.00	89492	33141	2901	56351	2.85R
60	5.00	37707	43781	7178	-6074	-0.33 X
66	2.00	115000	35183	4503	79817	4.11R
85	1.00	10000	32317	6867	-22317	-1.19 X
88	2.00	73529	30275	2962	43254	2.19R
90	5.00	84223	48688	6831	35535	1.90 X

R denotes an observation with a large standardized residual
X denotes an observation whose X value gives it large influence.

a) The plots indicate many problems with the model. The residuals are not normally distributed (from both the normality plot and the histogram). The I chart indicates that there are observations with unusual residuals and the plot of residuals vs. fitted values indicates that there might be a problem with equal variances.

Driver Habits

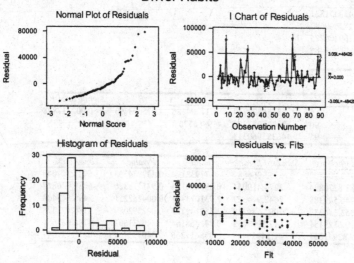

b) Correlations (Pearson)

```
          DRVRCNT  VEHCOUNT  NUM_KIDS  DTVCNT_H
VEHCOUNT   0.493
           0.000

NUM_KIDS   0.289    0.073
           0.006    0.496

DTVCNT_H   0.491    0.348     0.385
           0.000    0.001     0.000

DTVMILH    0.460    0.174     0.174     0.333
           0.000    0.101     0.102     0.001
```

None of the variables are very highly correlated. The significant correlations are not very big. In any case, the two variables in the model, number of vehicles and number of children have the lowest correlation. Multicollinearity might be the only problem this model does not have.

c) From the Minitab output, there are three observations with large X values and four with large residuals.

d) If we drop the observations with large X value influence, the model does not change much, it does not improve and now another observation is unusual.

Regression Analysis

```
The regression equation is
HHVMILES = 8167 + 5782 VEHCOUNT + 5758 NUM_KIDS
```

Predictor	Coef	StDev	T	P
Constant	8167	5691	1.44	0.155
VEHCOUNT	5782	2561	2.26	0.027
NUM_KIDS	5758	1995	2.89	0.005

```
S = 19695      R-Sq = 15.9%    R-Sq(adj) = 13.9%
```

Analysis of Variance

Source	DF	SS	MS	F	P
Regression	2	6180025295	3090012648	7.97	0.001
Residual Error	84	32582030223	387881312		
Total	86	38762055518			

Source	DF	Seq SS
VEHCOUNT	1	2948314761
NUM_KIDS	1	3231710535

Unusual Observations

Obs	VEHCOUNT	HHVMILES	Fit	StDev Fit	Residual	St Resid
8	2.00	101313	25488	2129	75825	3.87R
27	3.00	89492	31270	3189	58222	3.00R
50	2.00	14058	42761	6565	-28703	-1.55 X
65	2.00	115000	37003	4720	77997	4.08R
86	2.00	73529	31246	3071	42283	2.17R

```
R denotes an observation with a large standardized residual
X denotes an observation whose X value gives it large influence.
```

e) I would not use this model at all. It is not even a good model, even ignoring the problems. I would suggest a data transform of some kind and a new set of additional independent variables.

Chapter 13 Experimental Design and ANOVA - Solutions

Section 13.3 Exercises

13.1 a) The response variable is absorbency at time of failure. The factor is the method used.
b) There were 3 levels of the factor.
c) The ANOVA results are given below:

One-way ANOVA: Method 1, Method 2, Method 3

```
Analysis of Variance
Source     DF        SS        MS        F        P
Factor      2     102.70     51.35     5.57     0.005
Error     102     940.29      9.22
Total     104    1042.99
```

```
                                     Individual 95% CIs For Mean
                                     Based on Pooled StDev
Level      N      Mean      StDev   -+---------+---------+---------+-----
Method 1   35     4.800     4.727                  (-------*-------)
Method 2   35     3.314     1.641    (--------*-------)
Method 3   35     5.714     1.619                     (--------*-------)
                                    -+---------+---------+---------+-----
Pooled StDev =     3.036            2.4       3.6       4.8       6.0
```

At the 0.05 level of significance, there is a difference in average number of days on a breathing tube. It appears that Method 2 is different from Method 3.
d) The only conclusion you can draw is that there is a difference in the average. Without further analysis you cannot tell which means are different. Furthermore, you cannot draw any conclusions about one method being better than the others without studying additional variables such as patient health, long-term outcome, etc.

13.3 a) The response is grade. The factor is how homework was treated.
b) There are three levels of the factor.

c) One-way Analysis of Variance

```
Analysis of Variance
Source     DF        SS        MS        F        P
Factor      2     1135.2     567.6     6.58     0.002
Error      81     6988.6      86.3
Total      83     8123.8
```

```
                                     Individual 95% CIs For Mean
                                     Based on Pooled StDev
Level        N      Mean      StDev  -------+---------+---------+---------
No Homew    28    76.607     10.086       (------*------)
HomeworkNot 28    75.036      9.621   (------*------)
HomeworkCol 28    83.500      8.035               (------*------)
                                     -------+---------+---------+---------
Pooled StDev =     9.289             75.0      80.0      85.0
```

There is a significant difference in grades for the three methods. No homework and homework not collected are similar, and homework collected is different.
d) Based on the data, I would assign homework and collect it.

13.5 a) The response variable is distance traveled (carry). The factor is ball design.
b) There are four levels of the factor.

c) One-way Analysis of Variance

```
Analysis of Variance
Source     DF        SS        MS        F        P
Factor      3     1152.7     384.2     25.32    0.000
Error     140     2124.5      15.2
Total     143     3277.2
```

```
                                   Individual 95% CIs For Mean
                                   Based on Pooled StDev
Level        N      Mean    StDev  -+---------+---------+---------+-----
M1 model    36    257.42     2.36                              (---*---)
M2 model    36    256.78     2.29                            (---*---)
M3 model    36    250.19     5.40   (---*---)
M4 model    36    255.06     4.55                     (---*---)
                                   -+---------+---------+---------+-----
Pooled StDev =      3.90          249.0     252.0     255.0     258.0
```

There is a significant difference in carry for the four models. Model 1 is different from all of the others. Models 1 and 2 are similar and model 3 is similar to 2 but different from 1.

d) Based on the data I would recommend using either design 1 or 2 since they had the longest carry. The actual difference would depend on cost and other factors.

Section 13.4 Exercises

13.7   a) There does not appear to be a problem with the assumption of independence.

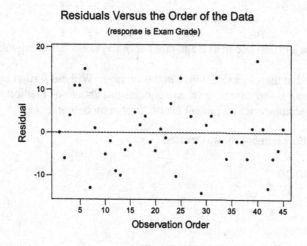

b) Looking at the histograms, the assumption of normality does not seem too extreme.

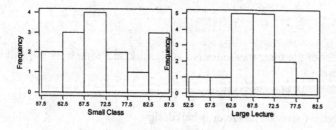

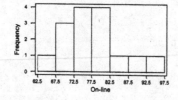

c) Looking at the boxplots and sample variances, the assumption of equal variances is a not a problem.

```
Level      N      Mean      StDev
Small Cl   15     72.000    8.552
Large Le   15     68.200    6.592
On-line    15     78.067    7.592
```

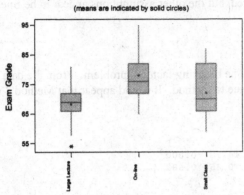

Boxplots of Exam Grade by Method of Teaching
(means are indicated by solid circles)

d) The assumptions of ANOVA are fine for this data.

13.9 a) There does appear to be some kind of pattern as you look at the data. Certainly, the variability is increasing, but there is no systematic variation that would indicate autocorrelation.

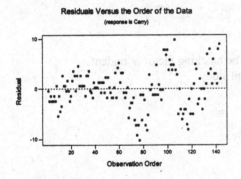

Residuals Versus the Order of the Data
(response is Carry)

b) From the histograms it would appear that the data are normally distributed.

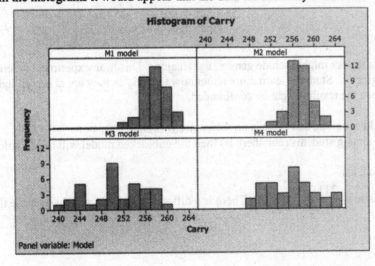

Histogram of Carry

c) Looking at the histograms and the variances, the assumption of equal variances is a problem. We saw this in the residual plot too.

```
Level       N     Mean    StDev   Variance
M1 model   36    257.42    2.36    5.57
M2 model   36    256.78    2.29    5.24
M3 model   36    250.19    5.40   29.16
M4 model   36    255.06    4.55   20.70
```

d) The assumption of equal variances is violated, but the other assumptions appear to be fine.

Section 13.5 Exercises

13.11   a) and b) The ANOVA table is given below. The blocking factor is problem. From the data we see that there is a significant difference in number of iterations due to method. It would appear that Method 3 uses the least number of iterations on the average.

```
Analysis of Variance for Iteration

Source    DF        SS        MS       F      P
Method     3    184.282    61.427   44.78  0.000
Problem   30     34.919     1.164    0.85  0.688
Error     90    123.468     1.372
Total    123    342.669
Means

Method   N   Iteration
1       31    3.8710
2       31    1.7419
3       31    0.6774
4       31    1.1613
```

13.13   a) The response variable is GPA.
b) The factor is circumstances under which you study. The blocking factor is student.
c) GPA is different for the circumstances under which you study.

```
Analysis of Variance for GPA

Source     DF       SS        MS       F      P
How         2    1.38464   0.69232  17.16  0.000
Students    9    3.17075   0.35231   8.73  0.000
Error      18    0.72629   0.04035
Total      29    5.28168

Means

How    N     GPA
1     10   3.2440
2     10   2.7200
3     10   2.9400
```

e) In a subsequent study, other factors might include gender, age, major...This is an experiment where randomization would matter greatly. Students learn more about how to study as they are at school longer. If the factors are studied in order, then the results might be confounded.

13.15   a) From the ANOVA we would conclude that there is no difference in weight over time.
b) If there is a lot of variation among students (and there is) then the unblocked model will not be able to detect the difference in mean weight.
c) The blocks would be the students.

d) From the output, the F statistic is $\dfrac{211.29}{2.79} = 75.73$. There is a difference in mean weight over the time periods.

Section 13.6 Exercises

13.17 The Minitab Output is given below:

Two-way Analysis of Variance

```
Analysis of Variance for Carry
Source        DF      SS        MS       F       P
Design         3   1152.72   384.24   70.69   0.000
Time           2    806.93   403.47   74.23   0.000
Interaction    6    600.07   100.01   18.40   0.000
Error        132    717.50     5.44
Total        143   3277.22
```

```
                       Individual 95% CI
Design    Mean    ---+---------+---------+---------+--------
1        257.42                                  (--*--)
2        256.78                               (--*--)
3        250.19      (--*--)
4        255.06                      (--*--)
                  ---+---------+---------+---------+--------
                250.00    252.50    255.00    257.50
```

```
                       Individual 95% CI
Time      Mean    -----+---------+---------+---------+------
Afternoo  257.58                                 (--*--)
Midday    255.19                       (--*--)
Morning   251.81    (--*--)
                  -----+---------+---------+---------+------
                  252.00    254.00    256.00    258.00
```

a) Since the p value is 0.000, there is a difference due to design.

b) There is a difference due to time of day.

c) There is an interaction effect. This means that the different designs act differently at different times of day. A look at a plot of the 12 means shows this effect.

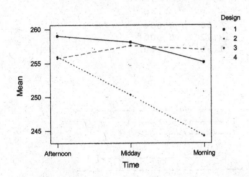

Interaction Plot - Data Means for Carry

d) To check normality we can look at a histogram and a normal plot. Both indicate that the assumption of normality is reasonable.

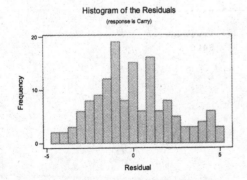

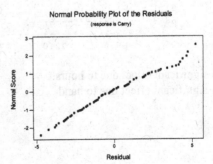

To look at the assumption of equal variances we can look at plots of the residuals vs. the fitted values. The assumption of equal variances seems reasonable.

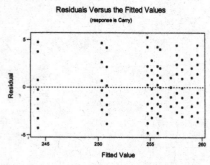

Residuals Versus the Fitted Values
(response is Carry)

A plot of the residuals vs. order indicates no problems with independence of the error terms.

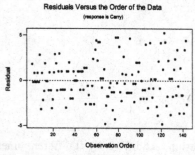

Residuals Versus the Order of the Data
(response is Carry)

13.19   The output is given below:

Two-way Analysis of Variance

Analysis of Variance for Time

Source	DF	SS	MS	F	P
Hours	3	6452	2151	4.66	0.004
Hand	1	6101	6101	13.22	0.000
Interaction	3	1874	625	1.35	0.259
Error	152	70130	461		
Total	159	84558			

```
                              Individual 95% CI
Hours          Mean      --------+---------+---------+---------+---
2 hours        77.1                        (-------*--------)
4 hours        77.1                        (-------*--------)
6 hours        64.4      (-------*--------)
Beginnin       81.2                          (--------*-------)
                         --------+---------+---------+---------+---
                            64.0      72.0      80.0      88.0
```

```
                              Individual 95% CI
Hand           Mean      ----+---------+---------+---------+-------
Left           81.1                        (-------*-------)
Right          68.8        (-------*-------)
                         ----+---------+---------+---------+-------
                            66.0      72.0      78.0      84.0
```

a) There is a significant effect due to hours.
b) There is a significant effect due to hand.

c) There is not a significant interaction effect.

d) The plot indicates that the time to complete the task does not vary consistently over time. After about two hours the times are the same, after four hours they are better with the right hand, worse with the left and after 6 hours people are best.

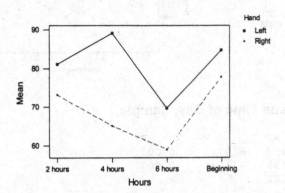

Interaction Plot - Data Means for Time

e) I have no idea. Perhaps expect people to start slowly? I certainly can't imagine having people warming up for 6 hours before they start work.

Chapter 13 Exercises
Learning It

13.21   a) It is a blocked design. The factor is type of connection, the blocking variable is day.

b) Two-way ANOVA: Failure Rate versus Connection, Day Number

```
Analysis of Variance for Failure
Source       DF        SS         MS         F        P
Connecti      2   0.0012066  0.0006033    96.90    0.000
Day Numb      9   0.0000716  0.0000080     1.28    0.313
Error        18   0.0001121  0.0000062
Total        29   0.0013903

                         Individual 95% CI
Connecti      Mean     -----+---------+---------+---------+------
Cable       0.03120                              (--*---)
DSL         0.01950      (--*--)
Phone di    0.03420                                    (--*---)
                         -----+---------+---------+---------+------
                        0.02000   0.02500   0.03000   0.03500
```

There is a significant difference due to connection. The failure rate is different (lower) for DSL than for the other two.

c) The plots are shown below. The normality assumption and the assumption of equal variances might be a problem here. A transform of the data might be appropriate.

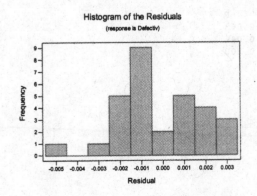

Histogram of the Residuals
(response is Defectiv)

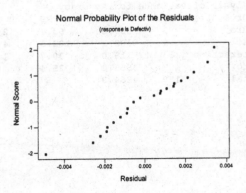

Normal Probability Plot of the Residuals
(response is Defectiv)

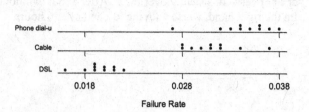

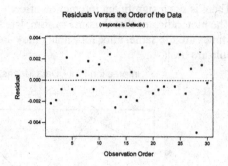

13.23   a) **Two-way ANOVA: Download versus Time of Day, Sample**

```
Analysis of Variance for Download
Source      DF       SS        MS       F        P
Time of      4    1.79540   0.44885  619.27   0.000
Sample      29   13.54149   0.46695  644.24   0.000
Error      116    0.08408   0.00072
Total      149   15.42097
```

```
                        Individual 95% CI
Time of        Mean    ----------+----------+----------+----------+-
10PM        14.7783                                          (*-)
3PM         14.4967    (*)
6AM         14.6400                            (*)
Midnight    14.7780                                          (*)
Noon        14.5900                  (*)
                       ----------+----------+----------+----------+-
                         14.5600   14.6400   14.7200   14.8000
```

b) There is still a difference due to time of day although the 10 p.m. downloads are the same as the midnight ones..

13.25   a) The average wait at the border is different before and after 9/11.

One-way Analysis of Variance

```
Analysis of Variance
Source     DF      SS       MS       F        P
Factor      1    240.7    240.7    16.99    0.015
Error       4     56.7     14.2
Total       5    297.3
```

```
                                 Individual 95% CIs For Mean
                                 Based on Pooled StDev
Level    N    Mean    StDev   -+---------+---------+---------+-----
After    3   25.667   3.512                    (---------*-------)
Before   3   13.000   4.000   (---------*-------)
                              -+---------+---------+---------+-----
Pooled StDev =   3.764         7.0      14.0      21.0      28.0
```

b) Other factors might be the day of week and time of day and nationality or country of origin of passport of the driver.

13.27   **Two-way Analysis of Variance**

```
Analysis of Variance for Accuracy
Source        DF       SS       MS       F       P
Hours          3     189.2     63.1     2.49    0.063
Hand           1     102.3    102.3     4.03    0.046
Interaction    3     119.0     39.7     1.56    0.200
Error        152    3854.4     25.4
Total        159    4264.9
```

```
                              Individual 95% CI
Hours        Mean    ---+---------+---------+---------+---------+-------
2 hours      91.99                        (---------*---------)
4 hours      91.15        (----------*---------)
6 hours      89.68    (----------*---------)
Beginnin     92.57                         (---------*---------)
                      ---+---------+---------+---------+---------+--------
                      88.50     90.00     91.50     93.00
```

```
                              Individual 95% CI
Hand         Mean    ------+---------+---------+---------+-----
Left         90.55   (----------*---------)
Right        92.15                    (----------*----------)
                     ------+---------+---------+---------+-----
                        90.00     91.00     92.00     93.00
```

a) and b) The output indicates that there is not a difference in accuracy due to time, but that there is one due to hand. This makes sense - people should use the hand they prefer. There is no significant interaction. There are no recommendations to make.

13.29   a)–c) There is a significant difference due to type of break and glue type. There is also an interaction.

Two-way Analysis of Variance

```
Analysis of Variance for Tensile
Source           DF       SS        MS        F        P
Type of Break     1     49665     49665    125.97    0.000
Glue              1    136960    136960    347.39    0.000
Interaction       1    144321    144321    366.07    0.000
Error             8      3154       394
Total            11    334101
```

```
                              Individual 95% CI
Type of Break   Mean    -+---------+---------+---------+---------+
Already        224.8                                   (---*----)
First           96.2    (----*----)
                        -+---------+---------+---------+---------+
                       80.0     120.0     160.0     200.0     240.0
```

```
                              Individual 95% CI
Glue            Mean    -----+---------+---------+---------+------
Regular        267.3                                   (---*--)
Gel             53.7    (--*--)
                        -----+---------+---------+---------+------
                          60.0     120.0     180.0     240.0
```

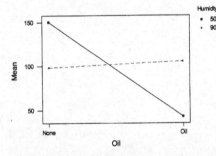

Interaction Plot - Data Means for Tensile1

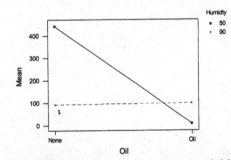

Interaction Plot - Data Means for Tensile 2

Looking at the interaction plots side by side we see that the Really Stick It company has much higher times than the Always Stick glue company, but it also has lower times. In regular type, the products are both high, and in gel type the Really Stick It is lower. If the regular type is used (due to price) then the two are comparable. If the gel is used, then the person who breaks things a lot should use the Always Stick Glue Company. In general a poor, clumsy person is probably best off with regular glue.

Thinking About It

13.31   It would have been very difficult to do a block design since there is no way to control the enrollment in the classes. If there is really a question about class to class variability, then the teacher could do a pre-test in each class to check that, or could use more than one test to compare the methods. As long as the same instructor is teaching all three sections, the sections should be comparable. This would be true unless one is at a really bad time like Saturday morning at 8 a.m., in which case it would be the LAST section students would register for and you might bias the results.

13.33   You should expect to find zero in the interval. If you fail to reject the null hypothesis you are saying that there is not a difference in the means, hence a confidence interval for the difference should contain 0.

13.35   a) No. It stands to reason that there will be an improvement in score as time goes on. You would not be able to tell if the change was due to defense or when the games occurred.
    b) A better design would be to assign each defense strategy to each time period (first, second and third thirds of the seasons). In addition, it would be best to randomly assign a strategy to a game in each time period.

13.37   The type of work would matter. Typing and word processing would be different from graphics or spreadsheet work. In addition, experience level at the particular task and with a computer might be factors.

Chapter 14 The Analysis of Qualitative Data - Solutions

NOTE: Answers may differ slightly due to rounding and use of graphing calculators or software vs. tables.

Section 14.2 Exercises

14.1   a) and b) Based on the table, the observed and expected frequencies seem to agree.

Age Group	Observed Frequency	Expected Frequency	(o-e)	$(o-e)^2/e$
Less than 18	6	6.8	-0.8	0.094118
18 - 19	118	75.3	42.7	24.21368
21 - 24	102	134.6	-32.6	7.895691
Greater Than 24	26	35.3	-9.3	2.450142
			Chi Square	34.65363

c) H_o: The distribution of students agrees with the theoretical distribution
   H_A: The distribution of students does not agree with the theoretical distribution
d) The degrees of freedom are 4 - 1 = 3. $\chi^2_{0.05, 3}$ = 7.815. Since the test statistic is 34.654, we reject H_o and conclude that the two distributions are not the same.

14.3   a) Because it is a success/failure event and you are sampling from a large population. We must assume that the probability that a person gets a seat remains constant, and that the trials are independent.
   b)–d) Note the collapsed cells because the expected frequencies are less than 5.

Number of Commuters in 15 Who Got Seats	Count	Expected	$(o-e)^2/e$
0	0	0.0	
1	0	0.0	
2	0	0.0	
3	0	0.0	
4	0	0.0	
5	0	0.0	
6	0	0.1	
7	0	0.3	
8	1	1.4	
9	3 (4)	4.3 (6.1)	0.7229508
10	10	10.3	0.008738
11	21	18.8	0.257447
12	31	25.0	1.44
13	20	23.1	0.436527
14	9	13.2	
15	5 (14)	3.5 (16.7)	0.436527
Total	100	Chi Square	3.281680

There are 6 - 1 = 5 degrees of freedom. The critical value is $\chi^2_{0.01, 5}$ = 15.086.
   e) Since the test statistic is less than the critical value we conclude that it is reasonable to assume that the data come from a binomial distribution with n = 15 and π = 0.80.

14.5   a) H_o: The data come from a binomial distribution with n = 10 and p= 0.80
   H_A: The data do not come from a binomial distribution with n = 10 and p= 0.80

b) Note the collapsed cells because of expected frequencies less than 5

Number of Women	Frequency	Expected	$(o-e)^2/e$
4	1	0.53	
5	5	2.29	
6	11(17)	7.74 (10.56)	3.93
7	19	17.69	0.10
8	27	26.58	0.01
9	18	23.58	1.32
10	7	9.42	0.62
		Chi Square	5.98

c) There are 5 - 1 = 4 degrees of freedom. The $\chi^2_{0.05, 4} = 9.488$ so we fail to reject H_o. It is reasonable to assume that the data are binomial with n = 10 and p = 0.80.

Section 14.3 Exercises

14.7   a) H_o: $p_1 = p_2 = p_3 = p_4 = p_5$
        H_A: at least one p is different
     b) The overall proportion of people who rate their phone service excellent is 1252/5452 = 0.22964
     c) **Chi-Square Test**
```
Expected counts are printed below observed counts

     Long Dis Local Ph  Power Cable TV Cellular  Total
  1     264      444     131      215      198   1252
      380.74   404.63  141.46   148.35   176.82

  2    1394     1318     485      431      572   4200
     1277.26  1357.37  474.54   497.65   593.18

Total  1658     1762     616      646      770   5452

Chi-Sq = 35.796 +  3.831 +  0.773 + 29.947 +  2.536 +
         10.671 +  1.142 +  0.230 +  8.927 +  0.756 = 94.610
DF = 4, P-Value = 0.000
```
     d) $\chi^2_{0.05, 4} = 9.488$
     e) There is a difference in the proportion who rate their cell phone service excellent by type of company.

14.9   a) H_o: $p_1 = p_2 = p_3 = p_4$
        H_A: at least one p is different
     b) The overall proportion of females is 56/201 = 0.279.
     c) **Chi-Square Test**
```
Expected counts are printed below observed counts

     Use it f  Use inf Heard Ab No exper   Total
  F     46        3        3        4        56
      45.69     4.18     2.23     3.90

  M    118       12        5       10       145
     118.31    10.82     5.77    10.10

Total  164       15        8       14       201

Chi-Sq =  0.002 +  0.333 +  0.267 +  0.003 +
          0.001 +  0.128 +  0.103 +  0.001 = 0.837
DF = 3, P-Value = 0.841
3 cells with expected counts less than 5.0
```
     d) At the 0.10 level of significance with 3 degrees of freedom, the critical value is 6.251.
     e) Since the p value is 0.841 (and the test statistic is 0.837) we can conclude gender and familiarity are not related.

Note: There are cells with expected frequencies below 5. Some cells should be combined (know somebody and no experience). Since the overall effect of small expected values is to inflate the Chi Square statistic and the value for this test is not beyond the critical value, the small expected frequencies does not present a problem.

Section 14.4 Exercises

14.11 a) H_o: Shift and type of defect are independent
   H_A: Shift and type of defect are not independent of each other.

b) Chi-Square Test

```
Expected counts are printed below observed counts

        Printing Rips/Tea   Size    Total
     1      55       60       85      200
           67.33    62.00    70.67

     2      58       63       79      200
           67.33    62.00    70.67

     3      89       63       48      200
           67.33    62.00    70.67

  Total    202      186      212      600

Chi-Sq =  2.259 +  0.065 +  2.907 +
          1.294 +  0.016 +  0.983 +
          6.972 +  0.016 +  7.270 = 21.782
DF = 4, P-Value = 0.000
```

c) There are 4 degrees of freedom.
d) At the 0.05 level of significance shift and type of defect are related.

14.13 a) H_o: Age and experience with online price comparison sitesare independent
   H_A: Age and experience are not independent.
b) There will be $(4 - 1)(3 - 1) = 6$ degrees of freedom.

c) Chi-Square Test

```
Expected counts are printed below observed counts
       Under 25 25 to 44 45 and U   Total
   Fre   11       51      101        163
         13.79    47.03   102.18

   Inf    2        2       11         15
          1.27     4.33     9.40

   Hea    2        2        4          8
          0.68     2.31     5.01

   No E   2        3       10         15
          1.27     4.33     9.40

  Total   17       58      126        201

Chi-Sq =  0.563 +  0.334 +  0.014 +
          0.422 +  1.252 +  0.271 +
          2.588 +  0.041 +  0.205 +
          0.422 +  0.408 +  0.038 = 6.558
DF = 6
* WARNING * 1 cells with expected counts less than 1.0
          * Chi-Square approximation probably invalid
6 cells with expected counts less than 5.0
```

d) The test statistic is 6.558.
e) The critical value is $\chi^2_{0.05, 6} = 12.592$. This would say that experience and age are independent.
f) The expected frequencies could be a problem. In this case, it does not affect the result, but it should be fixed.

Chapter 14 Exercises
Learning It

14.15   a) and b) The data might be normally distributed.

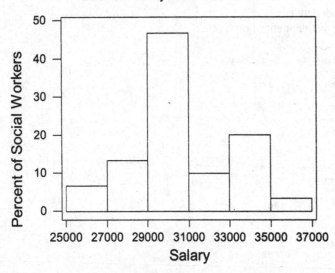

Salaries of Entry Level Social Workers

c) H_o: The data come from a N~(32500,930) distribution.

     H_A: The data do not come from a N~(32500,930) distribution.

d) and e) Since the test statistic is 0.5542 which is less than the critical value $\chi^2_{0.05,\,2} = 5.991$, we cannot reject H_o. The assumption of N~(32500,930) is not unreasonable.

From	To	Frequency	Expected	$(o-e)^2/e$
25000	27000	2	0	
27000	29000	4	0	
29000	31000	14	1.6	
31000	33000	3	19.55 (21.15)	0.1618
33000	35000	6	8.73 (8.84)	0.3930
35000	37000	1	0.11	
			Chi Square	

14.17   a) H_o: The proportion of people who take Prozac is the same for all regions of the US

     H_A: At least one region is different.

b) The overall proportion of adults who take Prozac 97/1175 = 8.3%.

c)

	New England	Middle Atlantic	East North Central	West North Central	South Atlantic	East South Atlantic	West South Central	Mountain	Pacific
Yes	5.0	13.1	16.9	7.8	18.7	6.2	9.4	7.3	12.6
No	55.0	145.9	188.1	86.2	208.3	68.8	104.6	80.7	140.4
Totals	60	159	205	94	227	75	114	88	153

Chi-square calculations
Variable B

		New England	Middle Atlantic	East North Central	West North Central	South Atlantic	East South Atlantic	West South Central	Mountain	Pacific
Variable A	Yes	1.8742	0.3443	0.0003	0.6466	2.0915	0.5283	1.2363	0.0096	4.6100
	No	0.1686	0.0310	0.0000	0.0582	0.1882	0.0475	0.1112	0.0009	0.4148

Chi-square test statistic =	12.3616
p-value =	0.1358

d) The value of the Chi Square test statistic is 12.3616.

e) The critical value is $\chi^2_{0.05, 8} = 15.507$. Since the test statistic is smaller than the critical value, we cannot reject H_o. There is no evidence that the proportion is different for different regions.

14.19 a) and b) From the histograms it looks like neither supplier meets the specs. Supplier A has too much variability and Supplier B is not quite symmetric.

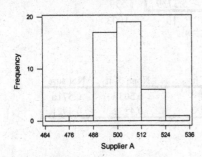 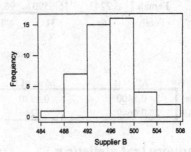

c) **Supplier A:** The critical value is $\chi^2_{0.05, 1} = 3.841$. Since the test statistic is less than the critical value we do not reject H_o and conclude that Supplier A is meeting the specifications.

From	To	Observed	Expected	$(o-e)^2/e$
464	476	1	0.0	
476	488	1	0.4	
488	500	17 (19)	22.1 (22.5)	0.5444
500	512	19	22.1 (22.5)	0.5444
512	524	6	0.4	
524	536	1 (26)	0.0	
		45	Chi Square	1.0889

Supplier B: The critical value is $\chi^2_{0.05, 3} = 7.815$. Since the test statistic is greater than the critical value we reject H_o and conclude that Supplier B is not meeting the specifications.

From	To	Observed	Expected	$(o-e)^2/e$
484	488	1	0.3	
488	492	7	2.1	
492	496	15 (23)	7.1 (9.5)	19.17174
496	500	16	13.0	0.709679
500	504	4	13.0	6.200458
504	508	2	7.1	
508	512	0	2.1	
512	516	0 (2)	0.3 (9.5)	5.923505
				32.00538

14.21 a) H_o: Gender and Amount of Shopping done on the Internet are Independent
H_A: Gender and Amount of Shopping done on the Internet are Dependent

b)

Expected frequencies
Variable B

		All of it	Most of it	Some of it	None of it	Not sure	Totals
Variable A	Male	2.7757	14.3410	82.8076	359.4497	4.6261	464
	Female	3.2243	16.6590	96.1924	417.5503	5.3739	539
	Totals	6	31	179	777	10	1003

c)

Chi-square calculations
Variable B

	All of it	Most of it	Some of it	None of it	Not sure
Male	0.5400	0.4930	1.5128	0.5033	0.5716
Female	0.4649	0.4244	1.3023	0.4332	0.4921

Chi-square test statistic =	6.7376
p-value =	0.1504

d) Since the p value is 0.1504, at the 0.05 level of significance there is no evidence that gender and amount of Internet shopping are dependent.

14.23 a) Looking at the data, it appears that the later months have higher frequencies than you would expect.

b) The critical value is $\chi^2_{0.05, 11} = 19.675$. Since the test statistic is not greater than the critical value, we cannot reject H_o. There is no reason to believe that the data were not randomly (uniformly) distributed.

Month	Number	Expected	$(o - e)^2/e$
January	13	15.25	0.331967
February	12	15.25	0.692623
March	9	15.25	2.561475
April	11	15.25	1.184426
May	14	15.25	0.102459
June	14	15.25	0.102459
July	14	15.25	0.102459
August	18	15.25	0.495902
September	19	15.25	0.922131
October	13	15.25	0.331967
November	21	15.25	2.168033
December	25	15.25	6.233607
		Chi Square	15.22951

14.25 a) The chi-square test is different because it tests whether the distribution is normal as well as the mean and the standard deviation. If you just test the mean and variance, you might have data that are skewed or uniformly distributed.

b) No, it means that it is not normal with that particular mean and that particular standard deviation.

c) Supplier A: $\overline{X} = 501.42$ and s = 10.33

Supplier B: $\overline{X} = 495.51$ and s = 4.44

d) Supplier A: We lose two more degrees of freedom because we calculated the parameters from the data. The critical value is $\chi^2_{0.05, 1} = 3.841$. Since the test statistic is less than the critical value, we can assume that the data are N~(501, 10.3).

From	To	Observed	Expected	(o -e)²/e
464	476	1	0.3	
476	488	1	4.3 (4.6)	1.508481
488	500	17	16.1	0.049791
500	512	19	17.8	0.078738
512	524	6	5.8	
524	536	1	0.6 (6.4)	0.051554
			Chi Square	1.688564

Supplier B: We have 1 degree of freedom here, so the critical value is the same. Since the Chi Square value is less than the critical value, we conclude that the data are N~(405,4,4).

From	To	Observed	Expected	(o -e)²/e
476	480	0	0.0	
480	484	0	0.3	
484	488	1	2.2	
488	492	7	8.6 (11.1)	0.887537
492	496	15	15.4	0.010378
496	500	16	12.7	0.857972
500	504	4	4.8	
504	508	2	0.8	
508	512	0	0.1 (5.7)	0.010597
			Chi Square	1.766484

b) Since we are using the data to estimate the mean and standard deviation, the number of degrees of freedom is much smaller.

c) In the first test we agreed that Supplier A met the specs and B did not. In the second test we are ignoring the specs and simply checking for normality.

14.27 a) and b) The boxplot shows several outliers. The data do not appear normally distributed.

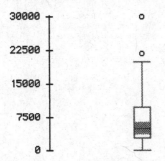

VEHMILES

c) After removing the outliers, the data still look very skewed.

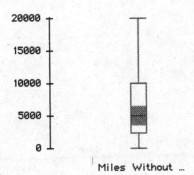

Miles Without ...

d) The mean is 6565.1 and the standard deviation is 5296.6.

e) Because the data are so skewed, the theoretical distribution will have negative values in it. You cannot ignore them. There are 5 - 1 - 2 = 2 degrees of freedom. The critical value is $\chi^2_{0.05,\,2} = 5.991$. Since the test statistic is greater than this value, we cannot say that the data are N~(6565,5297), and we never thought they were.

From	To	Observed	Expected	$(o-e)^2/e$
-10000	-5000	0	0.9	
-5000	0	0	6.1 (7.0)	7.043519
0	5000	26	18.2	3.31101
5000	10000	20	23.6	0.554036
10000	15000	11	13.4	0.422571
15000	20000	6	3.3	
20000	25000	3	0.4 (3.7)	0.117745
				11.44888

Chapter 15 Nonparametric Statistics - Solutions

Section 15.2 Exercises

15.1 a) From the dotplot, it appears the variability and shapes of the two populations are similar. Therefore, the hypothesis test is about the mean.

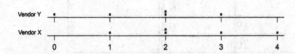

Number of Defective Cases

b) H_0: The mean number of defective cases from Vendor X is not greater than that for Vendor Y.
H_A: The mean number of defective cases from Vendor X is greater than that for Vendor Y.

c)

4	Vendor X	Vendor Y	Rank	Position
		0	1	1
		1	2.5	2
	1		2.5	3
		2	5.5	4
		2	5.5	5
	2		5.5	6
	2		5.5	7
		3	8.5	8
	3		8.5	9
	4		10	10
SUM	32	23		

d) At the 0.05 level of significance, the critical value for the upper tail test is 36 ($n_1=5$ and $n_2 = 5$). Therefore we fail to reject H_0. Conclude that there is not enough evidence to conclude that the mean number of defective cases from Vendor X is greater than that for Vendor Y.

15.3 a) From the dotplot, it appears the variability and shapes of the two populations are similar. Therefore, the hypothesis test is about the mean.

Daily Web Hits

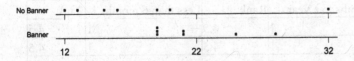

b) H_0: The mean number of daily hits for the banner site is not more than the no banner site.
H_A: The mean number of daily hits for the banner side is more than the no banner site.

c)

		Banner	No Banner	Rank	Position
			12	1	1
			13	2	2
			15	3	3
			16	4	4
		19		6.5	5
		19		6.5	6
		19		6.5	7
			19	6.5	8
			20	9	9
		21		10.5	10
		21		10.5	11
		25		12	12
		28		13	13
			32	14	14
	SUM	65.5	39.5		

d) The upper critical value for $n_1 = 7$ and $n_2 = 7$ is 66. Therefore there is not enough evidence to reject H_0. We can not conclude that the banner ads produce more hits at the .05 level of significance.

Section 15.3 Exercises

15.5   a) The dotplots for all three customer categories is shown below:

Customer Satisfaction

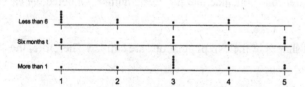

It appears that the variability is similar but the shapes are very different. Therefore the hypothesis test is about the population distributions.

b) H_0: There is no difference in customer satisfaction among the three groups.

   H_A: At least one is different

c)

More than a year	Rank	Six months to a year	Rank	Less than 6 months	Rank
1	4.5	1	4.5	1	4.5
2	10.5	1	4.5	1	4.5
3	17	2	10.5	1	4.5
3	17	3	17	1	4.5
3	17	3	17	1	4.5
3	17	3	17	2	10.5
3	17	4	23.5	2	10.5

4	23.5	5	28	3	17	
5	28	5	28	4	23.5	
5	28	5	28	4	23.5	
Rank sum	179.5		178		107.5	
Rank average	17.95		17.8		10.75	
Overall rank average		15.5				
SSA Contribution	60.025		52.9		225.625	
SSA	338.55					
SST	2125.5					
H	4.77841					

Rank	SST Contribution
4.5	121
10.5	25
17	2.25
17	2.25
17	2.25
17	2.25
17	2.25
23.5	64
28	156.25
28	156.25
4.5	121
4.5	121
10.5	25
17	2.25
17	2.25
17	2.25
23.5	64
28	156.25
28	156.25
28	156.25
4.5	121
4.5	121
4.5	121
4.5	121
4.5	121
10.5	25
10.5	25
17	2.25
23.5	64
23.5	64

d) The critical value is 5.991. Therefore since H=4.78 you fail to reject the null hypothesis. Conclude that there is no difference in satisfaction among the three groups.

15.7   a) It appears that the variability are similar but the shapes are different . Therefore the hypothesis test is about the distributions.

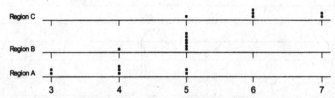

b) H_0: There is no difference in the use of public transportation among the three regions.
   H_A: At least one is different.

c)

	Region A	Rank	Region B	Rank	Region C	Rank
	3	1.5	4	4.5	5	11
	3	1.5	5	11	6	17
	4	4.5	5	11	6	17
	4	4.5	5	11	6	17
	4	4.5	5	11	7	19.5
	5	11	5	11	7	19.5
	5	11	5	11		
Rank Sum	38.5		70.5		101	
Average rank	5.5		10.07		16.83	
Overall Rank Average		10.5				
SSA Contribution	196		3.73		218.17	
SSA	417.9					
SST	598.8					
H	13.2600					

```
Minitab Output
Region      N     Median    Ave Rank          Z
A           7     4.000          5.5      -2.77
B           7     5.000         10.1      -0.24
C           6     6.000         16.8       3.13
Overall    20                   10.5
H = 11.91  DF = 2  P = 0.003
H = 13.27  DF = 2  P = 0.001 (adjusted for ties)
```

d) The critical value is 5.991 and since H is greater than this value, you reject the null hypothesis and conclude there is a difference in the use of public transportation by region.

Chapter 15 Exercises
Learning It!

15.9   a)

Dotplot for Low-High

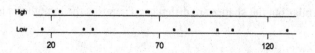

From the dotplot, it appears the variability and shapes of the two populations are similar. Therefore, the hypothesis test is about the mean.

b) H_0: The mean palatability is the same for both levels of liquid.

   H_A: The mean palatability is different for the two levels of liquid.

c)

	Low	High	Rank	Position
	16		1	1
		21	2	2
		24	3	3
	35		4	4
		39	5.	5
	39		5.5	6
		60	7	7
		64	8	8
		65	9	9
	77		10	10
	84		11	11
		86	12	12
		94	13	13
	97		14	14
	104		15	15
	129		16	16
SUM	765	595		

d) At the 0.05 level of significance, the critical values for the two tail test are 49 and 87 ($n_1=8$ and $n_2 = 8$). Therefore fail to reject H_0 because the sum of the ranks are in between the critical values and therefore not in the rejection region. Conclude that there is not enough evidence to conclude that the mean palatability is different for the two levels of liquid.

15.11   a) It appears that the variability and the shapes are not similar. Therefore the hypothesis test is about the distributions.

Dotplot for DSL-Satellite

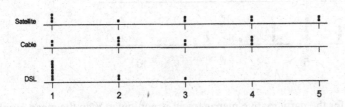

b) H_0: There is no difference in satisfaction for the three different high-speed Internet services.

   H_A: At least one is different

c)

DSL	Rank	Cable	Rank	Satellite	Rank
1	6.5	1	6.5	1	6.5
1	6.5	1	6.5	1	6.5
1	6.5	2	15.5	1	6.5
1	6.5	2	15.5	2	15.5
1	6.5	2	15.5	3	21
1	6.5	3	21	3	21
1	6.5	3	21	4	26
2	15.5	4	26	4	26
2	15.5	4	26	5	29.5
3	21	4	26	5	29.5
Rank Sum	**97.5**		**179.5**		**188**
Average rank	**9.75**		17.95		18.8
Overall Rank Average		15.5			
SSA Contribution	**330.625**		60.025		108.9
SSA	**499.55**				
SST	**2066.5**				
H	**7.01**				

```
Minitab Output

Type of    N    Median    Ave Rank         Z
Cable      10   2.500     18.0          1.08
DSL        10   1.000      9.8         -2.53
Satellit   10   3.000     18.8          1.45
Overall    30             15.5


H = 6.45  DF = 2  P = 0.040
H = 7.01  DF = 2  P = 0.030 (adjusted for ties)
```

d) The critical value is 5.991 and since H is greater than this value, you reject the null hypothesis and conclude there is a difference in the satisfaction for the three different services.

Thinking About It

15.13   a) The distributions look normally distributed.

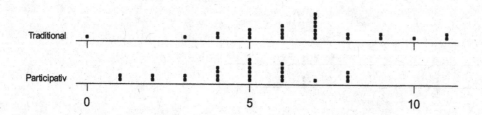

b) H_0: The mean number of sick days for the participative management is not lower than the mean number of sick days for the traditional management.

H_A: The mean number of sick days for the participative management is lower than the mean number of sick days for the traditional management.

```
From Minitab running the Wilcoxon Rank Sum (Mann-Whitney) Test

W = 493.5
Test of ETA1 = ETA2  vs  ETA1 < ETA2 is significant at 0.0027
```

```
The test is significant at 0.0025 (adjusted for ties)
```

Reject the null hypothesis and conclude that there is enough evidence to conclude that the mean number of sick days taken is less under a participative form of management.

c) Using a Student's t-test the results from Mintab are shown below:

Two-sample T for Participative vs Traditional

```
            N      Mean    StDev   SE Mean
Particip   25      4.76     2.01     0.40
Traditio   25      6.60     2.47     0.49

Difference = mu Participative - mu Traditional
Estimate for difference:  -1.840
95% upper bound for difference: -0.773
T-Test of difference = 0 (vs <): T-Value = -2.89  P-Value = 0.003  DF = 46
```

The p-value is very small (certainly less than .05) and so you reject the null hypothesis and conclude that there is enough evidence to conclude that the mean number of sick days taken is less under a participative form of management.

d) In this particular case, the difference in the mean number of sick days taken under the two management styles is great enough so that it doesn't matter which test you take. The distributions appear to be close enough to a normal distribution to allow the use of the Student's t-test which is always more desirable than the non-parametric test.

15.15   a) The graphs do not look normally distributed.

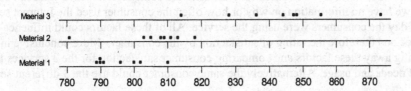

b) H_0: There is no difference in the mean amount of fluid that the diapers can absorb is not different for the three different materials.
H_A: At least one is different.

The results of the Kruskal-Wallis test are:

Material	N	Median	Ave Rank	Z
1	8	793.5	6.0	-3.18
2	8	808.5	11.1	-0.67
3	8	845.5	20.4	3.86
Overall	24		12.5	

H = 16.99 DF = 2 P = 0.000
H = 16.99 DF = 2 P = 0.000 (adjusted for ties)

Therefore, we reject the null and conclude that at least one of the materials leads to a different mean absorption of fluid.

c) Running the analysis as an ANOVA we get

One-way ANOVA: Amount of Fluid versus Material Type

```
Analysis of Variance for Amount o
Source     DF      SS      MS       F       P
Material    2    9864    4932   29.93   0.000
Error      21    3461     165
Total      23   13325

                              Individual 95% CIs For Mean
                              Based on Pooled StDev
Level       N    Mean   StDev  -------+---------+---------+---------
1           8  796.00    7.50  (----*----)
2           8  806.00   11.12     (----*----)
3           8  843.13   17.73                          (----*---)
                                 -------+---------+---------+---------
Pooled StDev =    12.84         800      820       840
```

The p value is very small (0.000) and so we reject the null hypothesis and conclude that at least one of the means is different.

d) Since the data does not look normally distributed, it is better to use the non-parametric test here.

15.17 a) In exercise 15.11 we concluded that there was a difference in customer satisfaction for the three Internet services considered. The service which had the lowest average rank (meaning the customers were more satisfied) was the DSL service. The average rank for the DSL service was found to be 9.75 compared to the other two rank averages of 17.95 and 18.8. Thus, not only was the average rank for the DSL service the lowest but it was also quite different from the other two services. Thus, based on this data I would recommend the DSL service.

b) In this study, the type of service was randomly assigned to each consumer. Although it is a good idea to use random assignment of services to users to avoid potential bias, there are some factors which were not considered in this study. Specifically, we have no information on why or how often the consumer used the Internet service. We don't know what time of day the consumers were using the service. All of these factors could influence the performance of the service and therefore the rating of satisfaction by the consumer. More conclusive information could be obtained by taking away these factors and comparing consumer satisfaction for the 3 services for consumers that have similar Internet needs and usage. Alternatively the same consumer could use the 3 different services for 3 different time periods.

Chapter 1 Extra Exercises - Solutions

Section 1.2 Exercises

1. A population is the entire group about which information is desired. The sample is a subset of a population on which measurements are made. They are the same when a census is taken.

3. a) The 12,000 working mothers in the area
   b) The 200 working mothers
   c) Sampling error

5. a) All babies this pediatrician delivers.
   b) Weight, gender, race/ethnicity, weeks of gestations, other possible
   c) Sample. The population keeps growing with each baby she delivers. It would not be possible to look at all babies she will ever deliver.

7. It is very difficult to measure the entire U.S. population, so a sample was most likely used to obtain this information.

9. a) All American adults
   b) A sample, the American population is too large and it is not feasible to poll everyone.

Section 1.3 Exercises

1. a) The percent of all American adults planning a vacation this summer
   b) Statistics, they were calculated from a sample.

3. a) The proportion of all college students who have missed at least one class due to alcohol or drug use.
   b) Statistic, it was calculated from a sample.

5. a. Statistic
   b) Parameter
   c) This is due to sampling error, we do not expect a statistic to be exactly equal to the parameter. The value varies from sample to sample.

7. a) Statistic
   b) Parameter
   c) This is due to sampling error, we do not expect a statistic to be exactly equal to the parameter. The value varies from sample to sample.

9. a) Parameter
   b) Statistic

Section 1.4 Exercises

1. Variability in the populations and size of the sample

3. a) N = 12,000
   b) n = 200

5. a) N = 270
   b) n = 40

7. a) N = 200 million
   b) n = 1002

9. Large variability in the population.

Section 1.5 Exercises

1. Yes, only the views of undergraduate students will be represented. Their views may differ from graduate students.

3. This is a volunteer response sample. Only people with strong opinions tend to reply.

5. They are compiled using a random number generator.

7. 9, 46, 32, 79, 57, 11, 50, 15, 34, 53

9. 9, 4, 6

Section 1.6 Exercises

1. a)   Quantitative
    b)   Qualitative
    c)   Quantitative

3. a.) Quantitative; continuous
    b)   Qualitative; nominal
    c)   Qualitative; nominal
    d)   Qualitative; ordinal
    e)   Quantitative; discrete

5. Ordinal qualitative

7. Continuous quantitative

9. Nominal qualitative

Section 1.10 Exercises

1. $\displaystyle\sum_{i=1}^{10} x_i = 288$

3. $\displaystyle\sum_{i=1}^{12} x_i = 1999$

5. $\displaystyle\sum_{i=1}^{11} x_i = 185$

7. $\displaystyle\sum_{i=1}^{8} x_i = 1520$

9. $\displaystyle\sum_{i=1}^{8} x_i = 216$

Chapter 2 Extra Exercises - Solutions

Section 2.2 Exercises

1.

Personal computer	320/1000	.320	32%
Pacemaker	260/1000	.260	26%
Wireless communication	180/1000	.180	18%
Television	240/1000	.240	24%

3.

Cups	Percent
0	6
1	10
2	37
3	18
4	29

5.
a) 10
b) 2
c) 5 to 7, 7 to 9, 9 to 11, 11 to 13, 13 to 15, 15 to 17, 17 to 19, 19 to 21, 21 to 23, and 23 to 26.
d)

class	frequency
5 to 7	1
7 to 9	5
9 to 11	6
11 to 13	16
13 to 15	20
15 to 17	16
17 to 19	15
19 to 21	10
21 to 23	6
23 to 26	5

Note that we chose to make the last class one unit longer to accommodate the data value 25. The sorted data is shown below.

```
5   7   7   8   8   8   9   9   9   9
10  10  11  11  11  11  11  11  12  12
12  12  12  12  12  12  12  12  13  13
13  13  13  13  13  13  13  13  13  13
13  14  14  14  14  14  14  14  15  15
15  15  15  15  15  15  15  15  15  16
16  16  16  16  17  17  17  17  17  17
17  18  18  18  18  18  18  18  18  19
19  19  19  20  20  20  20  20  20  21
21  21  21  22  22  23  23  24  24  25
```

7. a) The class $23 \leq x < 24$
b) The classes $20 \leq x < 21$ and $27 \leq x < 28$
c) The class $23 \leq x < 24$

9.

Injections	Percent
0	28%
1	46%
2	20%
3	5%
4	1%

Six from this group were eligible for the national study.

11.   d

13.   b

Section 2.3 Exercises

1.

3.

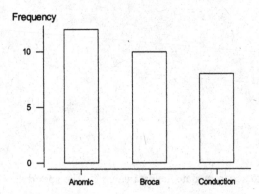

Type of Aphasia for thirty patients

5.

Type of Aphasia	Frequency
Conduction	24
Broca	18
Anomic	8

Conduction   (24/50)360 degrees = 172.8 degrees
Broca   (18/50)360 degrees = 129.6 degrees
Anomic   (8/50)360 degrees = 57.6 degrees

7.

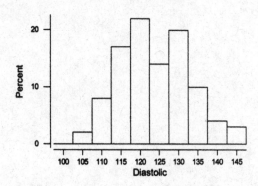

Diastolic blood pressures for Hypertensive patients

9.

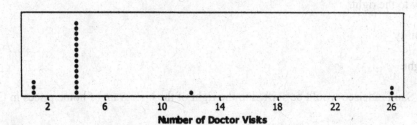

Dotplot of Doctor Visits

The value 4 occurs most often.

11.   b

13.   b

Section 2.4 Exercises

1.

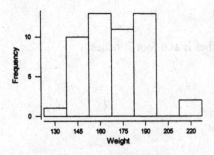

The center is close to 170 pounds. The data is symmetrically spread about the center.

3.

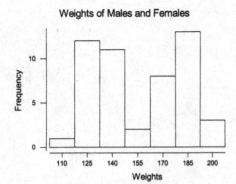

There is a mixture of two distributions here. One center is near 130 pounds and the other is near 180 pounds.

5. Most of the grooms are between 25 and 35 years of age. However a few are in their forties, fifties, sixties and seventies. This gives the distribution a skew to the right.

7. Machine 2 has the smaller variability.

9. LOS values are skewed to the right.

11. The histogram of home prices in San Francisco would be shifted to the right of the histogram of home prices in Omaha, Nebraska.

13. The histogram is symmetrical about a center of 80 and has a spread that goes from 50 to 110.

15.

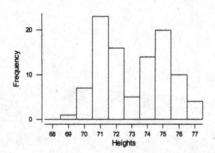

The distribution has two centers. One is at 5 foot 11 inches and the other is at 6 foot 3 inches.

Chapter Exercises

1.

Response	Frequency	Percent
No opinion	10	20%
Not at all	11	22%
Not too	8	16%
Somewhat	11	22%
Very	10	20%

3. Each class should be about 6.83 wide.

	Frequency
$14 \leq$ winning score < 21	7
$21 \leq$ winning score < 28	10
$28 \leq$ winning score < 35	7
$35 \leq$ winning score < 42	7
$42 \leq$ winning score < 49	2
$49 \leq$ winning score < 56	3

5.

Blood type	Percent	Angle
O	49	.49 x 360 degrees = 176.4 degrees
A	27	.27 x 360 degrees = 97.2 degrees
B	20	.20 x 360 degrees = 72 degrees
AB	4	.04 x 360 degrees = 14.4 degrees

7.

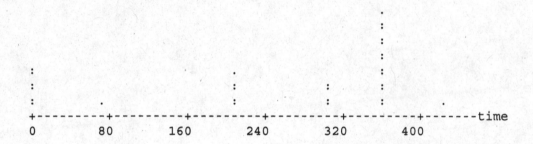

9. The distribution for the New Orleans daily high temperatures would be shifted to the right of the Seattle distribution.

Chapter 3 Extra Exercises - Solutions

Section 3.3 Exercises

1.

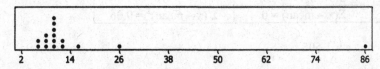

The mean is 15. The two extreme values, 25 and 86, pull the mean to the right of most of the data.

3.

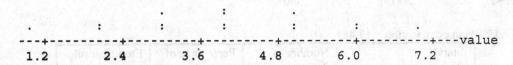

The mean, 4, is located at the center of the data.

5. Sample mean = 1340/100 = 13.4 credit hours

7.

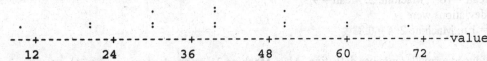

Mean = Median = Mode = 40.
In the case of a mound shaped symmetrical distribution, the mean, median, and mode are all equal.

9. The mean: sum = 70.97, mean = 70.97/40 = 1.77.
Trimmed mean: sum = 40.22, trimmed mean = 40.22/32 = 1.26
Comparing the two we see that they are different so some high data values are influencing the mean.

11. Σxf = 5028 Σf = 139 5028/139 = 36.2

13. The number 36 occurs most often. The mode is 36.

15. The median number of faxes is 22.

Section 3.4

1. Range = 6.5 – 3.4 = 3.1;   Mean = 5.2
Mean - .5Range = 3.7   Mean + .5Range = 6.7

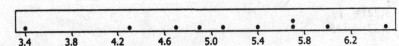

The range gives a good description of the data variability.

3.

x	(x − mean)	(x − mean)^2
1.5	1.5 − 3.4 = -1.9	3.61
3.4	3.4 − 3.4 = 0.0	0
5.4	5.4 − 3.4 = 2.0	4
4.2	4.2 − 3.4 = 0.8	.64
2.5	2.5 − 3.4 = - 0.9	.81
	Σ(x − mean) = 0	Σ (x − mean)^2 = 9.06

$s^2 = \Sigma$ (x − mean)^2/(n-1) = 9.06/4 = 2.265   s = 1.505

5.   Σx = 1533.9   Σx^2= 123448.8

$$s^2 = \frac{n\Sigma x^2 - (\Sigma x)^2}{n(n-1)} = \frac{20(123448.8) - 1533.9^2}{20(19)} = \frac{116126.79}{380} = 305.597 \quad S = 17.481$$

7.   Mean = 76.695; Standard deviation = 17.481

Interval	Number of data values	Percentage of data values	Empirical rule
59.214 to 94.176	13	65	68
41.733 to 111.657	19	95	95
24.252 to 129.138	20	100	99

9.   The means for the two machines were:
Machine 1: Mean = 10.1; Machine 2: Mean = 9.9
The standard deviations were :
Machine 1: s = 1.6; Machine 2: s = 0.83

Machine 2 produced a smaller number defective. Also, Machine 2 was much more consistent.

11.

$$s^2 = \frac{n\Sigma x^2 - (\Sigma x)^2}{n(n-1)} = \frac{41(4200) - (320)^2}{41(40)} = \frac{69800}{1640} = 42.56$$

13.   1900 − 150 = 1750 hours and 1900 + 150 = 2050 hours.

15.   z-score = (96 − 78)/12 = 18/12 = 1.5

Section 3.5 Exercises

1.   b = 51, e = 7, and n = 100. P = (51 + 7/2)/100 = .545 = 54.5%
Fifty-four and one half percent of the test scores were at or below hers.

3.   Lower inner fence = Q_1 − 1.5(IQR) = 47 − 1.5(22) = 14
Upper inner fence = Q_3 + 1.5(IQR) = 69 + 1.5(22) = 102
The lower whisker ends at 21 and the upper whisker ends at 96.

5.   Lower outer fence = Q_1 − 3(IQR) = 47 − 3(22) = -19
Upper outer fence = Q_3 + 3(IQR) = 69 + 3(22) = 135

7. Q_1 = (7 + 7)/2 = 7 Q_3 = (12 + 12)/2 = 12

9.   Lower outer fence = Q_1 − 3(IQR) = 7 − 3(5) = - 8
Upper outer fence = Q_3 + 3(IQR) = 12 + 3(5) = 27

11. $Q_1 = 54.375$ $Q_3 = 95.125$

13. Lower outer fence = $Q_1 - 3(IQR) = 54.375 - 3(40.75) = -67.875$
Upper outer fence = $Q_3 + 3(IQR) = 95.125 + 3(40.75) = 217.375$

15. $Q_1 = 47$ $Q_3 = 57$

17. Lower outer fence = $Q_1 - 3(IQR) = 47 - 3(10) = 17$
Upper outer fence = $Q_3 + 3(IQR) = 57 + 3(10) = 87$

19. $Q_1 = 12.5$ $Q_3 = 54.5$

21. Lower outer fence = $Q_1 - 3(IQR) = 12.5 - 3(42) = -113.5$
Upper outer fence = $Q_3 + 3(IQR) = 54.5 + 3(42) = 180.5$

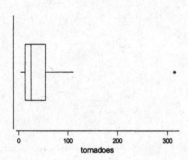

315 is a probable outlier.

Chapter 3 Exercises

1. Mean = 403.67 feet   Median = 410 feet   Mode = 410 feet (occurred 10 times)

3. Range = $488 - 320 = 168$ feet $\Sigma X^2 = 11{,}962{,}970$ $\Sigma X = 29{,}468$

$$s^2 = \frac{n\Sigma x^2 - (\Sigma x)^2}{n(n-1)} = \frac{73(11962970) - 29468^2}{73(72)} = \frac{4933786}{5256} = 938.69596 \qquad s = 30.638$$

5. $b = 65$, $e = 4$, and $n = 73$. $P = (65 + 4/2)/73 = .918 = 91.8\%$.
91.7% of the distances were at or below 440.

7. Lower inner fence = $Q_1 - 1.5(IQR) = 380 - 1.5(40) = 320$
Upper inner fence = $Q_3 + 1.5(IQR) = 420 + 1.5(40) = 480$
The lower whisker ends at 320 and the upper whisker ends at 454.

9.

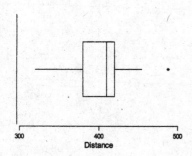

The 488 feet is a possible outlier.

Chapter 4 Extra Exercises - Solutions

Section 4.2 Exercises

1. Clustered bar chart with counties along the horizontal direction.

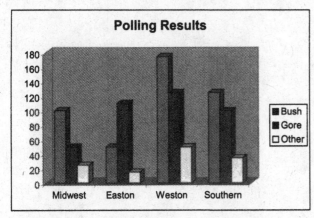

Bush has the greatest percentage lead in Midwest county, 57.1%. He is trailing in Easton county and has 28.56% of the voters.

3. Stacked bar chart with counties along the horizontal direction.

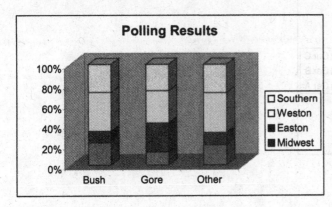

Bush has the majority in three of the four counties.

5. A = 40   B = 30   C = 45   D = 100

7.

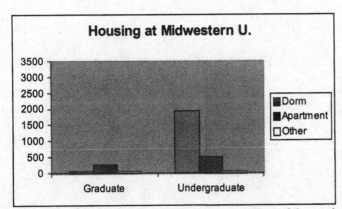

Most of the Undergraduates live in the dorm. Most of the graduate students live in apartments.

9.   C =D =E = 100%

11.   Clustered bar chart with rankings on the x-axis.

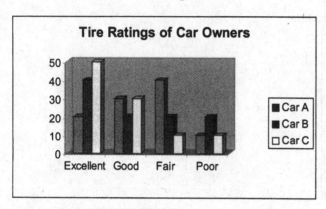

Overall the tire ratings were best with car C owners.

13.   Stacked bar chart with tire ratings on x-axis.

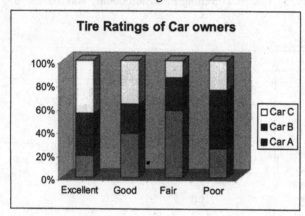

Section 4.3 Exercises

1.

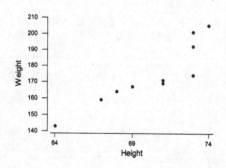

3. There is a non-linear relationship between the two variables.

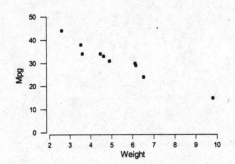

5. A negative linear relationship.

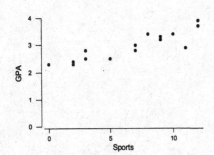

7. I would conclude that there is a negative linear relationship.

9. A direct linear relationship.

11.

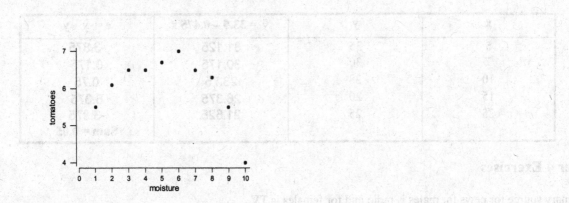

A negative linear relationship connects the two variables.

13. The sum is not zero because of round-off error.

x	y	$\hat{y} = 33.5 - 0.475x$	$e = \hat{y} - y$
5	35	31.125	-3.875
7	30	30.175	0.175
10	28	28.75	0.75
15	20	26.375	6.375
25	25	21.625	-3.375
			Sum = 0.05

Chapter 4 Exercises

1. Primary source for news for males is radio and for females is TV.

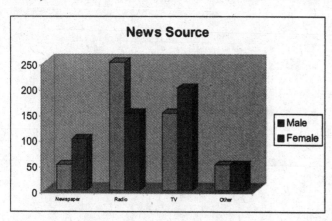

2.

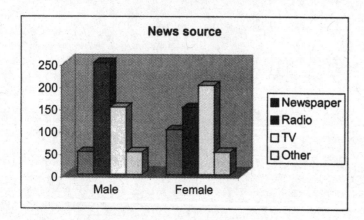

Females turn to Newspaper and TV for their news source, whereas males turn to Radio.

3.

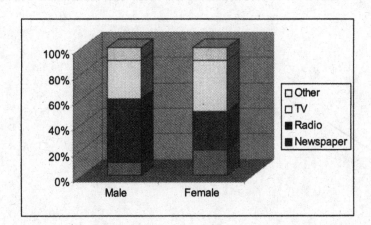

Fifty percent of the males choose radio as their primary source of news and forty percent of females choose TV as their primary source of news.

5.

	Newspaper	Radio	TV	Other	Total
Male	5%	25%	15%	5%	50%
Female	10%	15%	20%	5%	50%
Total	15%	40%	35%	10%	100%

7.

	Newspaper	Radio	TV	Other	Total(%)
Male	10%	50%	30%	10%	100%
Female	20%	30%	40%	10%	100%

9. $b = \dfrac{n\Sigma XY - \Sigma X \Sigma Y}{n\Sigma X^2 - (\Sigma X)^2} = \dfrac{5(3368750) - (11750)(1275)}{5(30687500) - (11750)^2} = \dfrac{1862500}{15375000} = 0.121$

$a = \dfrac{\Sigma Y}{n} - b\dfrac{\Sigma X}{n} = 255 - .121(2350) = -29.35$

$\hat{y} \quad = -29.35 + 0.121(3000) = \$333,650$

11. $b = \dfrac{n\Sigma XY - \Sigma X \Sigma Y}{n\Sigma X^2 - (\Sigma X)^2} = \dfrac{5(688.9) - (64.7)(56)}{5(889.23) - (64.7)^2} = \dfrac{-178.7}{260.06} = -0.687$

$a = \dfrac{\Sigma Y}{n} - b\dfrac{\Sigma X}{n} = 11.2 + 0.687(12.94) = 20.09$

$\hat{y} \quad = 20.09 - 0.687(12) = 11.846 = 12$ fires

13. $\hat{y} = -.1 + .7(3) = 2.0$. This is an example of interpolation. If the linear relationship is a strong one, the estimate will be a good one.

15.

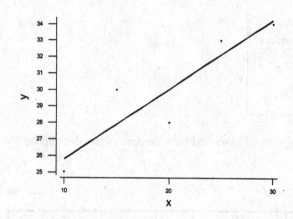

Chapter 5 Extra Exercises - Solutions

Section 5.2 Exercises

1. The sample space is represented as the 36 points in the following figure. Each outcome in the sample space is equally likely to occur.

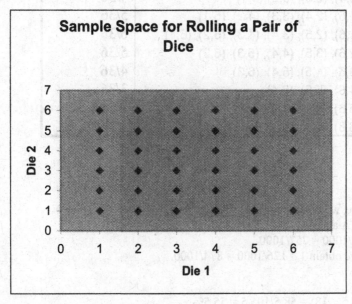

3. The sample space consists of one of the following outcomes. Each outcome in the sample space is equally likely to occur.

A♣	2♣	3♣	4♣	5♣	6♣	7♣	8♣	9♣	10♣	J♣	Q♣	K♣
A♦	2♦	3♦	4♦	5♦	6♦	7♦	8♦	9♦	10♦	J♦	Q♦	K♦
A♥	2♥	3♥	4♥	5♥	6♥	7♥	8♥	9♥	10♥	J♥	Q♥	K♥
A♠	2♠	3♠	4♠	5♠	6♠	7♠	8♠	9♠	10♠	J♠	Q♠	K♠

5. Referring to the sample space for problem 1 of this section, we see that there are 36 equally likely outcomes possible for this experiment. Let A be the event that the sum of the faces is not equal to 2. The A′ is that the sum of the faces is equal to 2. Since A′ consists of only one outcome, (1, 1) we know that P(A′) = 1/36. Therefore P(A) = 1 - P(A′) = 35/36.

7. a) .26 + .25 = .51
   b) 1 - .19 = .81

9. Not using the complement: .32 + .20 + .11 = .63
   Using the complement: 1 - .12 - .25 = 1 - .37 = .63

11.

sum	Outcome	Probability
2	(1,1)	1/36
3	(1,2), (2,1)	2/36
4	(1,3), (2,2), (3,1)	3/36
5	(1,4), (2,3), (3,2),(4,1)	4/36
6	(1,5), (2,4), (3,3), (4,2), (5,1)	5/36
7	(1,6), (2,5), (3,4), (4,3), (5,2), (6,1)	6/36
8	(2,6), (3,5), (4,4), (5,3), (6,2)	5/36
9	(3,6), (4,5), (5,4), (6,3)	4/36
10	(4,6), (5,5), (6,4)	3/36
11	(5,6), (6,5)	2/36
12	(6,6)	1/36

Section 5.3 Exercises

1. a) Using the general addition rule, we obtain
   P(A or B) = P(A) + P(B) – P(A and B)
   = 500/1000 + 446/1000 – 189/1000 = 757/1000.
   b) Using the complement rule, we obtain 1 – 126/1000 = 874/1000.
   c) 336/1000

3. a) (40.5 + 18)/(10.9 + 36.1 + 40.5 + 18) = 58.5/105.5 = 55.5%
   b) 18/(10.9 + 36.1 + 40.5 + 18) – 15.9/(10.6 + 31 + 34.4 + 15.9)
   = 18/105.5 – 15.9/91.9 = 0.171 – 0.173 = -0.002 = -0.2%.
   The percent decreased from 1990 to 2000.

5. a) Using the complementary property, 1 – 85/377 = 292/377 = 0.775.
   b) Using the general addition rule, 239/377 + 208/377 – 132/377 = 315/377 = 0.836.

7. The event A consists of the 13 clubs and the event B consists of the 13 hearts. Since A and B are mutually exclusive, the simple addition rule applies. P(A or B) = P(A) + P(B) or P(A or B) = 13/52 + 13/52 = 26/52.

9. The probability of an odd number is 6/12, since an odd number occurs for the outcomes 1H, 3H, 5H, 1T, 3T, 5T. The probability of head occurring is 6/12 since a head occurs for the outcomes 1H, 2H, 3H, 4H, 5H, 6H. Now an odd number and a head occurs for the outcomes 1H, 3H, 5H. Using the general addition rule, the probability is
6/12 +6/12 – 3/12 = 9/12.

11. The cards that are black and face cards are the following. The probability of the event A and B is 6/52.

J♣	Q♣	K♣
J♠	Q♠	K♠

13. Let A be the event that it was hit to left field and B be the event that it was hit in the fifth inning.

   P(A or B) = P(A) + P(B) – P(A and B) = 3/73 + 9/73 – 2/73 = 10/73.

Section 5.4 Exercises

1. A consists of 12 face cards and B consists of the 13 clubs. P(A|B) = P(A and B)/P(B) = (3/52)/(13/52) = 3/13; P(B|A) = P(A and B)/P(A) = (3/52)/(12/52) = 3/12

3. A = {HHT, HTH, THH, HHH} and B = {HTH, THH, HHT}
    P(B|A) = P(A and B)/P(A) = (3/8)/(4/8) = 3/4

5. A = {(1,1), (1,2), (1,3), (1,4), (1,5), (1,6)}
    B = {(1,1), (1,2), (2,1), (1,3), (2,2), (3,1), (1,4), (2,2), (3,1), (4,1), (1,5), (2,4), (3,3), (4,2), (5,1)}
    A and B = {(1,1), (1,2), (1,3), (1,4), (1,5)}
    P(A) = 6/36 = 1/6; P(A|B) = P(A and B)/P(B) = (5/36)/(15/36) = 5/15 = 1/3.
    Therefore, A and B are dependent.

7.

Chat Room	19 or less	20 to 50	More than 50	Total
Yes	134	75	34	243
No	34	84	125	243
Total	168	159	159	486

P(19 or less | Have been in Chat room) = (134/486)/(243/486) = 134/243 = 0.551

9. A and B independent means P(A and B) = P(A)P(B) = (.4)(.3) = .12.
    Using the general addition rule of probability, we have
    P(A or B) = P(A) + P(B) – P(A and B) = .4 + .3 - .12 = 0.58.

Chapter 5 Exercises

1. There are 104 outcomes in the sample space. The sample space may be visualized as follows:

Head with each of the following

A♣	2♣	3♣	4♣	5♣	6♣	7♣	8♣	9♣	10♣	J♣	Q♣	K♣
A♦	2♦	3♦	4♦	5♦	6♦	7♦	8♦	9♦	10♦	J♦	Q♦	K♦
A♥	2♥	3♥	4♥	5♥	6♥	7♥	8♥	9♥	10♥	J♥	Q♥	K♥
A♠	2♠	3♠	4♠	5♠	6♠	7♠	8♠	9♠	10♠	J♠	Q♠	K♠

Or

Tail with each of the following

A♣	2♣	3♣	4♣	5♣	6♣	7♣	8♣	9♣	10♣	J♣	Q♣	K♣
A♦	2♦	3♦	4♦	5♦	6♦	7♦	8♦	9♦	10♦	J♦	Q♦	K♦
A♥	2♥	3♥	4♥	5♥	6♥	7♥	8♥	9♥	10♥	J♥	Q♥	K♥
A♠	2♠	3♠	4♠	5♠	6♠	7♠	8♠	9♠	10♠	J♠	Q♠	K♠

3.   There are 4 favorable outcomes and 104 outcomes in the sample space. The probability is 4/104.

The four favorable outcomes are:

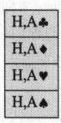

| H,A♣ |
| H,A♦ |
| H,A♥ |
| H,A♠ |

5.   Event A is as follows:

H,K♣	T,K♣
H, K♦	T,K♦
H,K♥	T,K♥
H,K♠	T,K♠

Event B is as follows.

H,Q♣	T,Q♣
H,Q♦	T,Q♦
H,Q♥	T,Q♥
H,Q♠	T,Q♠

Since A and B are mutually exclusive, we use the simple addition rule
P(A or B) = P(A) + P(B) = 8/104 + 8/104 = 16/104.

7.   Using the complementary rule, 100% - 5% = 95%.

9.   230/16230 = 0.014

Chapter 6 Extra Exercises - Solutions

Section 6.2 Exercises

1. a) Yes, $0 \le p(x) \le 1$, and $\Sigma\, p(x) = 1$
    b) No, not all $0 \le p(x)$.
    c) No, $\Sigma\, p(x) > 1$

3.

5.

x	Outcome	Probability
2	(1,1)	1/36
3	(1,2), (2,1)	2/36
4	(1,3), (2,2), (3,1)	3/36
5	(1,4), (2,3), (3,2),(4,1)	4/36
6	(1,5), (2,4), (3,3), (4,2), (5,1)	5/36
7	(1,6), (2,5), (3,4), (4,3), (5,2), (6,1)	6/36
8	(2,6), (3,5), (4,4), (5,3), (6,2)	5/36
9	(3,6), (4,5), (5,4), (6,3)	4/36
10	(4,6), (5,5), (6,4)	3/36
11	(5,6), (6,5)	2/36
12	(6,6)	1/36

x	2	3	4	5	6	7	8	9	10	11	12
p(x)	.028	.056	.083	.111	.139	.167	.139	.111	.083	.056	.028

7. $1/2 + 1/4 + 1/8 + 1/16 + \ldots$ is an infinite geometric series with first term $a = 1/2$ and common ratio $r = 1/2$. The sum of such a series is $a/(1-r) = 1/2/(1-1/2) = 1$.

9. Such a distribution is said to be skewed to the right.

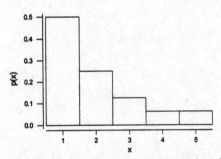

11. Such a distribution is said to be skewed to the left.

13. Such a distribution is said to be mound or bell shaped.

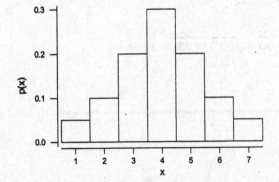

15. two

Section 6.3 Exercises

1. 1. Each experiment has a fixed number of trials, 15.
2. Success is being left-handed failure is not being left-handed.
3. p and q remain the same from trial to trial.
4. If the 15 are randomly chosen, the trials are independent.
5. The random variable is the number of left-handers that are in the 15.

3. mean = np = 15(.15) = 2.25   standard deviation = $\sqrt{np(1-p)} = \sqrt{15(.15)(.85)} = 1.383$
The z-score is (10 – 2.25)/1.383 = 5.6. Ten may be unusual.

5. mean = np = 10(.10) = 1 standard deviation = $\sqrt{np(1-p)} = \sqrt{10(.1)(.9)} = 0.949$.
The z-score is (4 - 1)/0.949 = 3.16. If the 10% figure is correct, they may have an unusual sample.

7. Skewed to the right.

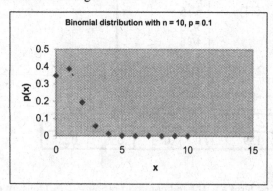

9. Skewed to the left.

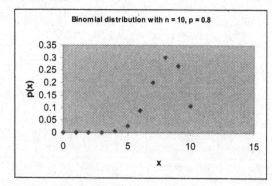

11. $p(0) = \dfrac{3!}{0!3!}(.5)^0(.5)^3 = .125$ $p(1) = \dfrac{3!}{1!2!}(.5)^1(.5)^2 = .375$

$p(2) = \dfrac{3!}{2!1!}(.5)^2(.5)^1 = .375$ $p(3) = \dfrac{3!}{3!0!}(.5)^3(.5)^0 = .125$

13. mean = np = 20(.75) = 15   standard deviation = $\sqrt{np(1-p)} = \sqrt{20(.75)(.25)} = 1.936$

15. mean = np = 20(1/6) = 3.33   standard deviation = $\sqrt{np(1-p)} = \sqrt{20(.167)(.833)} = 1.668$

z = (5 – 3.33)/1.668 = 1.00 Not too unusual to get a z-score of 1.

Section 6.4 Exercises

1.   P(X< a)

3.   P(X < a OR X > b)

5.   P(X > a)

Section 6.5 Exercises

1.   P(Z < 1.23) = 0.8907

3.   P(-1.56 < Z < 1.87) = P(Z < 1.87) – P(Z < -1.56) = 0.9693 – 0.0594 = 0.9099

5.   We are looking for P(X < 42.5). The Z value that corresponds to 42.5 is Z = (42.5 – 47.5)/2.5 = - 2.00. P(X < 42.5) is equal to P(Z < -2.00). From the tables this is equal to 0.0228. 2.28% of the women will have onset before 42.5.

7.   We are looking for P(3 < X < 8). The Z value corresponding to X = 3 is (3 – 5)/1.2 = -1.67 and the Z value corresponding to X = 8 is (8 – 5)/1.2 = 2.50.   P(3 < X < 8) = P(-1.67 < Z < 2.50). P(-1.67 < Z < 2.50) = P(Z < 2.50) – P(Z < -1.67) P(-1.67 < Z < 2.50) = 0.9938 – 0.0475 = 0.9463 or 94.63% spend between 3 and 8 hours in the chat room per week.

9.   We are looking for P(X > 700). The Z value corresponding to X = 700 is (700 – 550)/75 = 2.00. P(X > 700) = P(Z > 2.00) = 1 - P(Z < 2.00) = 1 - .9772 = 0.0228. On 2.28% on the days the number of e-mails exceeds 700.

11.   The Z value with 90% of the area to the left of it is 1.28. 1.28 = (x – 500)/100 Solving for x, we find 628.

13.   The binomial has a mean equal to np = 50 a standard deviation equal to $\sqrt{npq} = \sqrt{100(.5)(.5)} = 5$. Center the normal curve at 50 and choose the one that has a standard deviation equal to 5.

15.   We are looking for P(X > 200). The Z value that corresponds to 200 is Z = (200 – 180)/25 = 0.80. P(X > 200) is equal to P(Z > 0.80). From the tables this is equal to 1 - P(Z < 0.80) = 1 - .7881 = 0.2119; 21.2% need to be treated.

Chapter 6 Exercises

1.

x	0	1	2	3	4
p(x)	0.0625	0.2500	0.3750	0.2500	0.0625

3.   P(X = 0) = (48/52)(47/51) = 2256/2652 = 0.851
P(X = 1) = (4/52)(48/51) + (48/52)(4/51) = 192/2652 + 192/2652 = 384/2652 = 0.145
P(X = 2) = (4/52)(3/51) = 12/2652 = .004

x	0	1	2
P(x)	0.851	0.145	0.004

5.   The center of the normal curve would be at μ = 25(.5) = 12.5.
The standard deviation would be $\sqrt{25(.5)(.5)} = 2.5$.

7. We are looking for P(75 < X < 125). The z value corresponding to X = 75 is   Z = (75 – 94.50)/15.50 = -1.26 and the Z value corresponding to X = 125 is (125 – 94.50)/15.50 = 1.97. P(75 < X < 125) = P(-1.26 < Z < 1.97) = 0.9756 - 0.1038 = 0.8718.

9. Assume p = 0.75 and n = 25. Find p(0) + p(1) + ….+ p(12) = 0.0034. This casts doubt on the 75% figure.   Also check the Z value for X = 12, with mean = 25(.75) = 18.75 and standard deviation = $\sqrt{25(.75)(.25)}$ = 2.165. Z = (12 – 18.75)/2.165 = -3.11. The number 12 seems to be unusually small if p = 0.75 is correct.

Chapter 7 Extra Exercises - Solutions

Section 7.2 Exercises

1. point estimate

3. mean, median, trimmed mean, or mode

5. a) 2.68
   b) 1.8
7. a) 159.675
   b) 180

Section 7.3 Exercises

Output for 1 and 2

```
Summary Statistics:
Variable    n      Mean        Variance     Std. Dev.    Median
TwinA       19     5.4901314   0.9011559    0.9492923    5.6875
TwinB       19     5.6710525   1.0728481    1.0357838    5.5

Variable    Range   Min      Max       Q1        Q3
TwinA       3.875   3.5      7.375     4.6875    6.125
TwinB       3.625   3.9375   7.5625    4.9375    6.125
```

1. a) 5.49 lbs
   b) 5.67 lbs
   c) −0.18 lbs
   d) The difference is due to taking a sample rather then looking at the entire population. Values will vary from sample to sample.

Output for 3 and 4

```
Variable     N   N*    Mean   SE Mean   StDev   Minimum     Q1   Median      Q3
Control      23   0   41.52    3.58     17.15     10.00   28.00   42.00   54.00
Treatment    21   0   51.48    2.40     11.01     24.00   43.50   53.00   58.50

Variable    Maximum
Control      85.00
Treatment    71.00
```

3. a) 41.5 points
   b) 51.5 points
   c) 10 points
   d) No, the sample difference is 10, which suggests the population difference is above 0.

5. a) The proportion of all voters who will vote for the Republican candidate.
   b) 0.45

7. a) The proportion of all people who prefer the taste of the name brand soft drink and the proportion of all people who prefer the taste of the store brand soft drink.
   b) Name brand: 0.605   Store brand: 0.395

Section 7.4 Exercises

1. A point estimator should be unbiased, consistent, and efficient.

3. A point estimator is consistent if it yields a value close to the unknown parameter as the sample size increases.

5. a)

	Mean	Median
Sample 1	1.76	1.9
Sample 2	2.12	2
Sample 3	1.94	2
Sample 4	1.88	1.7
Sample 5	1.96	1.9

    b) Average = 1.932, Std Dev = 0.1308
    c) Average = 1.9 Std Dev = 0.1225
    d) The mean
    e) The median for these samples although it would usually be the mean.

Section 7.5 Exercises

1. sampling distribution

3. a) Normal
    b) 8.2 years
    c) 0.183 years

5. a) Normal
    b) 9 minutes
    c) 0.237 minutes

7. a) Normal
    b) 7.3 lbs
    c) 0.1 lb

Section 7.6 Exercises

1. If the distribution of the individual values is normal or if the sample size is sufficiently large, 30 or greater.

3. a) 0.367 years
    b) 0.183 years
    c) 0.110 years
    d) As the sample size increases, the standard error decreases.

5. a) 0.625 inches
    b) 0.357 inches
    c) 0.208 inches
    d) As the sample size increases, the standard error decreases.

7. a) 9 min; 0.237 min
    b) 9 min; 0.190 min
    c) Both populations have the same mean, but the one with the smaller standard deviation has a smaller standard error.

Section 7.7 Exercises

1. a) 7.65 to 8.75 years
    b) Z = -4.92
    c) Yes, the Z-score is very small, which indicates that this sample mean is very unusual for the claimed population mean.

3. a) 8.526 to 9.474 min
    b) Z = -1.69
    c) No, the Z-score indicates that this sample mean is within two standard deviations of the mean, which is not so unusual.

5. a) 7.0 to 7.6 lbs.
    b) Z = 2.00
    c) No, the Z-score indicates that this sample mean is two standard deviations above the mean, which is not so unusual.

7. a) 71.904 to 78.096
    b) Z = 4.84
    c) Yes, the Z-score is very large, which indicates that this sample mean is very unusual for the claimed population mean.

Section 7.8 Exercises

1. 262.77 to 273.23 days

3. 169.4 to 174.6 lbs.

5. 7.1 to 9.3 lbs.

7. 62.26 to 65.8 years

Section 7.9 Exercises

1. The Z-score formula involves the population standard deviation and the t-score formula replaced the population standard deviation with the sample population.

3. df = n – 1

5. Practice reading t-table:
    a. 2.492
    b. 1.725
    c. 2.056
    d. 2.831

Section 7.10 Exercises

1. 174.83 to 205.16 mg/dL

3. $31.68 to $38.49

5. 22.98 to 31.02 min

7. 1.92 to 9.08 shots

Section 7.11 Exercises

1. 13.8% to 21.2%

3. a) 8.7% to 13.3%
   b) Yes, since this 90% confidence interval is entirely below 20%.
   c) If the students in the sample lied, then the true value of the population parameter is actually higher than the estimated values from the confidence interval.

5. 0.13 to 0.24

7. a) 10.3% to 14.2%
   b) No, since the 95% confidence interval is entirely above 10%. The true percent is probability higher than 10%.

9. 17.4% to 34.6% of cancer patients to go into remission.

Section 7.12 Exercises

1. 40 pregnancies

3. 35 subjects

5. 20 households

7. 423 traffic fatalities

9. 196 movies

Chapter 8 Extra Exercises - Solutions

Section 8.3 Exercises

1. B

3. C

5. G

7. C

9. H

Section 8.4 Exercises

1. 1) State the null and alternative hypotheses
   2) Decide test procedure, statistic, and α
   3) Calculate the test statistic and p-value
   4) Make a decision
   5) Interpret the results

3. H_0: $p \geq 0.20$, H_a: $p < 0.20$

5. H_0: $\mu = 6$ oz, H_a: $\mu \neq 6$ oz

7. When the p-value is less than or equal to 0.10.

9. When the p-value is less than or equal to 0.05.

11. p-value

Section 8.5 Exercises

1. Z

3. T

Section 8.6 Exercises

1. a) H_0: $\mu = 266$, H_a: $\mu \neq 266$
   b) $\alpha = 0.05$, $z_{cutoff} = \pm 1.96$, $z = (268 - 266)/(4/\sqrt{36}) = 3.00$
      Therefore, the calculated value of z is not in the critical region.
   c) p-value $= 2(1 - 0.9987) = 0.0026$
   d) Reject the null hypothesis and conclude that the mean length of all pregnancies at this hospital is not similar to the overall average.

3. a) H_0: $\mu = 7$, H_a: $\mu \neq 7$
   b) $\alpha = 0.10$, $z_{cutoff} = \pm 1.645$, $z = (6.8 - 7)/(1.6/\sqrt{64}) = -1.00$
      Therefore, the calculated value of z is not in the critical region.
   c) p-value $= 2(0.1587) = 0.3174$
   d) We fail to reject the null hypothesis and conclude that the mean birth weight in this state does not differ from 7 lbs.

5. a) H_0: $\mu = 170$, H_a: $\mu \neq 170$
   b) $\alpha = 0.05$, $z_{cutoff} = \pm 1.96$, $z = (175 - 170)/(10/\sqrt{40}) = 3.16$
      Therefore, the calculated value of z is in the critical region.
   c) p-value = $2(1 - 0.9992) = 0.0016$
   d) We reject the null hypothesis and conclude that the mean weight of adult males differs from 170 lbs.

7. a) H_0: $\mu = 6.20$, H_a: $\mu \neq 6.20$
   b) $\alpha = 0.05$, $z_{cutoff} = \pm 1.96$, $z = (6.65 - 6.20)/(1.90/\sqrt{81}) = 2.13$
      Therefore, the calculated value of z is in the critical region.
   c) p-value = $2(1 - 0.9834) = 0.0332$
   d) We reject the null hypothesis and conclude that the mean bill at the owner's franchise is not the same as before the promotional campaign.

9. a) H_0: $\mu = 1$, H_a: $\mu \neq 1$
   b) $\alpha = 0.01$, $z_{cutoff} = \pm 2.57$, $z = (0.97 - 1)/(0.10/\sqrt{49}) = -2.10$ Therefore, the calculated value of z is not in the critical region.
   c) p-value = $2(0.0179) = 0.0358$
   d) We fail to reject the null hypothesis and conclude that the mean weight for all packages of ground beef does not differ from one pound.

Section 8.7 Exercises

1. a) No error
   b) Type I error

3. a) Type II error
   b) Type I error

5. a) Type II error
   b) Type I error

7. a) The mail order company's advertisements will be misleading in that their competitor actually does deliver the product within 4 days.
   b) The mail order company will think their competitor delivers the products as claimed, when they actually do not. This information will not be advertised.
   c) 0.01, since a type I error is more serious than a type II error.

9. a) The consumer magazine will assert that the manufacturer is incorrect about the stopping distance when the stopping distance is actually 25 ft, as claimed.
   b) The consumer magazine will assert that the manufacturer is correct about the stopping distance, when the car actually stops in more than 25 ft.
   c) 0.05, this is the standard value and both errors are about equally as serious.

Section 8.8 Exercises

1. H_0: $\mu \geq 24$ oz, H_a: $\mu < 24$ oz

3. H_0: $\mu = 266$ days, H_a: $\mu \neq 266$ days

5. H_0: $\mu = 50$ lbs, H_a: $\mu \neq 50$ lbs

7. H_0: $\mu \geq 5$ lbs, H_a: $\mu < 5$ lbs

9. H_0: $\mu = 30$ years, H_a: $\mu \neq 30$ years

Section 8.9 Exercises

1. a) $H_0: \mu \geq 24$ oz, $H_a: \mu < 24$ oz
    b) $\alpha = 0.10$, $z_{cutoff} = -1.28$, $z = (23.94 - 24)/(0.13/\sqrt{101}) = -4.64$
        Therefore, the calculated value of z is in the critical region.
    c) p-value $= 0.0000017$
    d) We reject the null hypothesis and conclude that the mean amount of cereal in a package is less than 24 oz.

3. a) $H_0: \mu \geq 9.4$ lbs, $H_a: \mu < 9.4$ lbs
    b) $\alpha = 0.05$, $z_{cutoff} = -1.645$, $z = (8.2 - 9.4)/(4/\sqrt{50}) = -2.12$
        Therefore, the calculated value of z is in the critical region.
    c) p-value $= 0.0170$
    d) We reject the null hypothesis and conclude that the mean weight of paper discarded by households is lower since the recycling program has been initiated.

5. a) $H_0: \mu \leq 8$ hours, $H_a: \mu > 8$ hours
    b) $\alpha = 0.05$, $z_{cutoff} = 1.645$, $z = (8.6 - 8)/(1.5/\sqrt{55}) = 2.97$ Therefore, the calculated value of z is in the critical region.
    c) p-value $= 0.0015$
    d) We reject the null hypothesis and conclude that the mean time the pain reliever controls muscle pain is longer than 8 hours.

7. a) $H_0: \mu \geq 480$ min, $H_a: \mu < 480$ min
    e) $\alpha = 0.05$, $z_{cutoff} = -2.05$, $z = (527.5 - 480)/(90/\sqrt{36}) = 3.17$
        Therefore, the calculated value of z is in the critical region.
    f) p-value $= 0.0008$
    g) We reject the null hypothesis and conclude that the mean time spent sleeping by children, age 3-5, is greater than 480 min (8 hours).

9. a) $H_0: \mu \leq 60$ years, $H_a: \mu > 60$ years
    b) $\alpha = 0.05$, $z_{cutoff} = 1.645$, $z = (64.03 - 60)/(13.53/\sqrt{225}) = 4.47$
        Therefore, the calculated value of z is in the critical region.
    c) p-value $= 0.0000$ (0.000004)
    d) We reject the null hypothesis and conclude that the mean age of billionaires is over 60 years of age.

Chapter 9 Extra Exercises - Solutions

Section 9.2 Exercises

1. a) H_0: $\mu \leq 180$
        H_A: $\mu > 180$
    b) sample mean = 190, sample standard deviation = 18.13
        $t = (190 - 180)/(18.13/\sqrt{8}) = 1.56$, p-value = 0.081
    c) At $\alpha = 0.01$, $t_{\text{cutoff}} = 2.998$ (upper-tail test, df = 7) Do not reject the null hypothesis.
    d) There is not enough evidence to conclude that the mean cholesterol reading of men age 60 and older is higher than 180 mg/dL.

3. a) H_0: $\mu = 40$
        H_A: $\mu \neq 40$
    b) sample mean = 35.08, sample standard deviation = 6.57
        $t = (35.08 - 40)/(6.57/\sqrt{12}) = -2.59$, p-value = 0.025
    c) At $\alpha = 0.05$, $t_{\text{cutoff}} = +/- 2.201$ (two-tail test, df = 11) Reject the null hypothesis.
    d) There is enough evidence to conclude that the mean price of a ticket concert differs from $40.

5. a) H_0: $\mu \geq 30$
        H_A: $\mu < 30$
    b) sample mean = 27, sample standard deviation = 6
        $t = (27 - 30)/(6/\sqrt{8}) = -1.41$, p-value = 0.100
    c) At $\alpha = 0.05$, $t_{\text{cutoff}} = -1.895$ (lower-tail test, df = 7) Do not reject the null hypothesis.
    d) There is not enough evidence to conclude that the mean time in which racers currently finish is less than 30 min.

7. a) H_0: $\mu = 7$
        H_A: $\mu \neq 7$
    b) sample mean = 5.5, sample standard deviation = 4.28
        $t = (5.5 - 7)/(4.28/\sqrt{8}) = -0.99$, p-value = 0.354
    c) At $\alpha = 0.05$, $t_{\text{cutoff}} = +/-2.365$ (two-tail test, df = 7) Do not reject the null hypothesis.
    d) There is not enough evidence to conclude that the mean number of flu shots given by doctors in this community differs from the national average of 7.

9. a) H_0: $\mu = 30$ min
        H_A: $\mu \neq 30$ min
    b) sample mean = 28.84 min, sample standard deviation = 5.145 min
        $t = (28.84 - 30.00)/(5.145/\sqrt{25}) = -1.127$, p-value = 0.2708
    c) At $\alpha = 0.05$, $t_{\text{cutoff}} = +/- 2.064$ (two-tail test, df = 24) Do not reject the null hypothesis.
    d) Based on the sample evidence, the foreman can report to the owner of the company that there is no difference between the lunch breaks being taken by the employees and the 30 minutes being allowed.

11. a) H_0: $\mu \leq 312$ lb
        H_A: $\mu > 312$ lb
    b) sample mean = 355.95 lb, sample standard deviation = 56.04 lb
        $t = (355.95 - 312)/(56.04/\sqrt{22}) = 3.68$, p-value = 0.0006
    c) At $\alpha = 0.01$, $t_{\text{cutoff}} = 2.518$ (upper-tail test, df = 21) Reject the null hypothesis.
    d) Based on the sample evidence, Coach Juarez can conclude that the average bench press now surpasses the 312-lb mean obtained by last year's team.

Section 9.3 Exercises

1. a) H_0: $\sigma^2 \leq 0.36$
        H_A: $\sigma^2 > 0.36$
    b) sample variance = 0.49
        $\chi^2 = ((30 - 1)*0.49)/0.36 = 14.21/0.36 = 39.47$
    c) At $\alpha = 0.05$, χ^2 upper = 42.6 (df = 29, upper-tail test) Do not reject the null hypothesis.
    d) There is not enough evidence to conclude that the variance in birth weights of premature infants is greater then the variance in birth weights of full-term infants.

3. a) H_0: $\sigma^2 \geq 73.96$
        H_A: $\sigma^2 < 73.96$
    b) sample variance = 40.96
        $\chi^2 = ((27 - 1)*40.96)/73.96 = 1064.96/73.96 = 14.40$
    c) At $\alpha = 0.05$, χ^2 lower = 15.379 (df = 26, lower-tail test) Reject the null hypothesis.
    d) There is enough evidence to conclude that the standard deviation in wait time has been reduced to less than 8.6 min.

5. a) H_0: $\sigma^2 = 0.50$
        H_A: $\sigma^2 \neq 0.50$
    b. sample variance = 0.65
        $\chi^2 = ((25 - 1)*0.65)/0.50 = 15.6/0.50 = 31.2$
    c. At $\alpha = 0.01$, χ^2 upper = 45.6, lower = 9.886 (df = 24, two-tail test) Do not reject the null hypothesis.
    d. There is not enough evidence to conclude that the variance in weight loss of these dieters differs from 0.50.

7. a) H_0: $\sigma^2 \leq 100$
        H_A: $\sigma^2 > 100$
    b) sample variance = 328.857
        $\chi^2 = ((8 - 1)*328.857)/100 = 2300.999/100 = 23.01$
    c) At $\alpha = 0.05$, χ^2 upper = 14.07 (df = 7, upper-tail test) Reject the null hypothesis.
    d) There is enough evidence to conclude that the standard deviation in cholesterol levels for men over 60 is higher than 10 mg/dL.

9. a) H_0: $\sigma^2 = 4$
        H_A: $\sigma^2 \neq 4$
    b) sample variance = 13.167
        $\chi^2 = ((13 - 1)*13.167)/4 = 158.004/4 = 39.501$
    c) At $\alpha = 0.10$, χ^2 upper = 21.0, lower = 5.226, (df = 29, two-tail test) Reject the null hypothesis.
    d) There is enough evidence to conclude that the variance in the amount of sugar contained in breakfast cereals differs from 4.

Section 9.4 Exercises

1. a) H_0: p ≥ 0.20
        H_A: p < 0.20
    b) sample proportion = 0.175, $Z = -1.25$, p value = 0.1056
    c) At $\alpha = 0.05$, $Z_{cutoff} = -1.645$ (lower-tail test) Do not reject the null hypothesis.
    d) There is not sufficient evidence to conclude that fewer than 20% of residents in this community live below poverty.

3. a) $H_0: p \geq 0.20$
   $H_A: p < 0.20$
   b) sample proportion = 0.11, $Z = -5.03$, p value = 0.0000 (2.44×10^{-7})
   c) At $\alpha = 0.05$, $Z_{cutoff} = -1.645$ (lower-tail test). Reject the null hypothesis.
   d) There is sufficient evidence to conclude that fewer that 20% of students at this college have used illegal drugs in the past month.
   b. The sample would not be truly representative of the population, which would mean we could not trust the results of this test.

5. a) $H_0: p \geq 0.25$
   $H_A: p < 0.25$
   b) sample proportion = 0.185, $Z = -2.12$, p value = 0.0169
   b) At $\alpha = 0.05$, $Z_{cutoff} = -1.645$ (lower-tail test). Reject the null hypothesis.
   c) There is sufficient evidence to conclude that the percentage of alcohol related traffic fatalities in this officer's jurisdiction is less than 25%.

7. a) $H_0: p = 0.10$
   $H_A: p \neq 0.10$
   b) sample proportion = 0.12, $Z = 2.07$, p value = 0.0385
   c) At $\alpha = 0.05$, $Z_{cutoff} = 1.96$ (two-tail test). Reject the null hypothesis.
   d) There is sufficient evidence to conclude that the percent of left-handed people differs from 10%.

9. a) $H_0: p = 0.70$
   $H_A: p \neq 0.70$
   b) sample proportion = 0.73, $Z = 0.891$, p value = 0.373
   c) At $\alpha = 0.05$, $Z_{cutoff} = +/- 1.96$ (two-tail test). Do not reject the null hypothesis.
   d) Based on the sample evidence, you can conclude that there is no difference between the proportion of households owning a working television set in your city and the nation as a whole.

Chapter 10 Extra Exercises - Solutions

Section 10.4 Exercises

1. $H_0: \mu_1 = \mu_2$ $H_a: \mu_1 \neq \mu_2$
    $z = 5.983$, p-value = 0.0000 ($2.196*10^{-9}$)
    At $\alpha = 0.05$, $z_{cutoff} = +/- 1.96$ (two-tail test)
    Reject the null hypothesis. There is enough evidence to conclude that the mean shopping time at the two stores are not the same.

3. $H_0: \mu_T \geq \mu_{Th}$ $H_a: \mu_T < \mu_{Th}$
    $z = -1.546$, p-value = 0.0610
    At $\alpha = 0.05$, $z_{cutoff} = -1.645$ (one-tail lower test)
    Do not reject the null hypothesis. There is not enough evidence to conclude that the mean rating for the Tuesday night class is lower than the mean rating for the Thursday night class.

5. $H_0: \mu_{cust} \leq \mu_{spec}$ $H_a: \mu_{cust} > \mu_{spec}$
    $z = 5.234$, p-value = 0.000
    At $\alpha = 0.05$, $z_{cutoff} = 1.645$ (one-tail upper test)
    Reject the null hypothesis. There is enough evidence to conclude the mean selling price for custom built homes is higher than mean selling price for speculation homes.

7. $H_0: \mu_A = \mu_B$ $H_a: \mu_A \neq \mu_B$
    $z = -6.795$, p-value = 0.000
    At $\alpha = 0.01$, $z_{cutoff} = +/- 2.575$ (two-tail test)
    Reject the null hypothesis. There is enough evidence to conclude that the mean time spent with the technical staff differs for company A and company B.

9. $H_0: \mu_A = \mu_B$ $H_a: \mu_A \neq \mu_B$
    $z = 1.230$, p-value = 0.2188
    At $\alpha = 0.05$, $z_{cutoff} = +/- 1.96$ (two-tail test)
    Do not reject the null hypothesis. There is not enough evidence to conclude that the mean distance for the golf balls differs between brand A and brand B.

Section 10.5 Exercises

1. $H_0: \mu_1 = \mu_2$ $H_a: \mu_1 \neq \mu_2$
    $t = 20.40$, p-value = 0.000
    At $\alpha = 0.01$, df = 23, $t_{cutoff} = +/- 2.807$ (two-tail test)
    Reject the null hypothesis. There is enough evidence to conclude that the mean measurement of the speed of light differs between the two methods, that used in 1879 and that used in 1882.

3. $H_0: \mu_{BigTen} = \mu_{Others}$ $H_a: \mu_{BigTen} \neq \mu_{Others}$
    $t = -1.51$, p-value = 0.142
    At $\alpha = 0.10$, df = 32, $t_{cutoff} = +/- 1.691$ (two-tail test)
    Do not reject the null hypothesis. There is not enough evidence to conclude that the mean salary for professors at the Big Ten universities differs from the mean salary for professors at other universities.

5. $H_0: \mu_{before} = \mu_{after}$ $H_a: \mu_{before} \neq \mu_{after}$
    $t = 2.86$, p-value = 0.003
    At $\alpha = 0.01$, df = 39, $t_{cutoff} = +/- 2.750$ (two-tail test)
    Reject the null hypothesis. There is enough evidence to conclude that the mean blood pressure for this patient changed while taking the medication.

7. H_0: $\mu_P = \mu_T$ H_a: $\mu_P \neq \mu_T$
    t = 0.802, p-value = 0.4261
    At α = 0.10, df = 50, t_{cutoff} = 1.676 (two-tail test)
    Do not reject the null hypothesis. There is not enough evidence to conclude the mean octane level differs between the two gas stations.

9. H_0: $\mu_{male} = \mu_{female}$ H_a: $\mu_{male} \neq \mu_{female}$
    t = 0.29, p-value = 0.773
    At α = 0.05, df = 29, t_{cutoff} = +/- 2.045 (two-tail test)
    Do not reject the null hypothesis. There is not enough evidence to conclude that the mean time spent studying on week nights differs between males and females.

Section 10.7 Exercises

1. a) H_0: $\mu_d \leq 0$ H_a: μ_d 0
        t = 3.71, p-value = 0.0005
        At α = 0.05, df = 24, t_{cutoff} = 1.711 (one-tail upper test)
    b) Reject the null hypothesis. There is enough evidence to conclude the mean score on the real estate IQ test is higher after taking the course.

3. H_0: $\mu_d = 0$ H_a: $\mu_d \neq 0$
    t = 5.65, p-value = 0.000
    At α = 0.05, df = 15, t_{cutoff} = +/- 2.131 (two-tail test)
    Reject the null hypothesis. There is enough evidence to conclude that the mean treadwear   measure differs between the two methods.

5. H_0: $\mu_d \leq 0$ H_a: $\mu_d > 0$
    t = 2.46, p-value = 0.012
    At α = 0.01, df = 18, t_{cutoff} = 2.552 (one-tail upper test)
    Do not reject the null hypothesis. There is not enough evidence to conclude the mean participation rate of women in the work force was higher in 1972 than in 1968.

7. H_0: $\mu_d \leq 0$ H_a: $\mu_d > 0$
    t = 8.94, p-value = 0.000
    At α = 0.01, df = 19, t_{cutoff} = +/- 2.539 (one-tail upper test)
    Reject the null hypothesis. There is enough evidence to conclude the mean refusal rate in mortgage lending is higher for high income minorities than for high income whites.

9. H_0: $\mu_d \geq 0$ H_a: $\mu_d < 0$
    t = -2.67, p-value = 0.007
    At α = 0.05, df = 19, t_{cutoff} = -1.729 (one-tail lower test)
    Reject the null hypothesis. There is enough evidence to conclude that, in 1960, the mean percent employed in agriculture was lower than the mean percent employed in industry.

Section 10.8 Exercises

1. H_0: $p_{before} \geq p_{after}$ H_a: $p_{before} < p_{after}$
    z = -2.77, p-value = 0.003
    At α = 0.01, z_{cutoff} = -2.33 (one-tail lower test)
    Reject the null hypothesis. There is sufficient evidence to conclude the proportion of passengers checking bags increased after the new system was installed.

3. H_0: $p_{good} \geq p_{fair/poor}$ H_a: $p_{good} < p_{fair/poor}$
    z = -2.19, p-value = 0.014
    At α = 0.05, z_{cutoff} = -1.645 (one-tail lower test)

Reject the null hypothesis. There is sufficient evidence to conclude the proportion of overweight people in "good" health is lower than the proportion of overweight people in "fair" or "poor" health.

5. H_0: $p_{placebo} \geq p_{drug}$ H_a: $p_{placebo} < p_{drug}$
   z = - 0.61, p-value = 0.272
   At α = 0.10, z_{cutoff} = -1.28 (one-tail lower test)
   Do not reject the null hypothesis. There is not sufficient evidence to conclude the proportion of people who experience drowsiness is lower for those taking the placebo than for those taking the experimental drug.

7. H_0: $p_{women} \leq p_{men}$ H_a: $p_{women} > p_{men}$
   z = 12.61, p-value = 0.000
   At α = 0.05, z_{cutoff} = 1.645 (one-tail upper test)
   Reject the null hypothesis. There is sufficient evidence to conclude the proportion of single women with children living in poverty is higher than the proportion of single men with children living in poverty.

9. H_0: $p_{males} = p_{females}$ H_a: $p_{males} \neq p_{females}$
   z = 1.19, p-value = 0.235
   At α = 0.10, z_{cutoff} = +/-1.645 (two-tail test)
   Do not reject the null hypothesis. There is not sufficient evidence to conclude the proportion of males preferring the name brand cola differs from the proportion of females preferring the name brand cola.

Section 10.9 Exercises

1. H_0: $\sigma^2_A \leq \sigma^2_B$ H_a: $\sigma^2_A > \sigma^2_B$
   F = 2.224, p-value = 0.007
   At α = 0.05, df_{num} = 40, df_{den} = 40, F_{cutoff} = 1.693 (one-tail test upper)
   Reject the null hypothesis. There is enough evidence to conclude that the variance in the distance traveled by brand A golf balls is greater than that of brand B.

3. H_0: $\sigma^2_{1882} = \sigma^2_{1879}$ H_a: $\sigma^2_{1882} \neq \sigma^2_{1879}$
   F = 7.968, p-value = 0.000
   At α = 0.01, df_{num} = 22, df_{den} = 99, F_{cutoff} = 2.39 (two-tail test)
   Reject the null hypothesis. There is enough evidence to conclude that the variance of the two methods to measure the speed of light differ.

5. H_0: $\sigma^2_{Oakland} \leq \sigma^2_{SanFran}$ H_a: $\sigma^2_{Oakland} > \sigma^2_{SanFran}$
   F = 3.157, p-value = 0.0491
   At α = 0.05, df_{num} = 10, df_{den} = 9, F_{cutoff} = 3.137 (one-tail test upper)
   Reject the null hypothesis. There is enough evidence to conclude that the variance in attendance increased after the team moved to Oakland.

7. H_0: $\sigma^2_{after} = \sigma^2_{before}$ H_a: $\sigma^2_{after} \neq \sigma^2_{before}$
   F = 1.283, p-value = 0.5924
   At α = 0.05, df_{num} = 19, df_{den} = 19, F_{cutoff} = 2.53 (two-tail test)
   Do not reject the null hypothesis. There is not enough evidence to conclude that the variance in the number of daily disruptions changed after the teacher began using remedial interventions.

9. H_0: $\sigma^2_{males} = \sigma^2_{females}$ H_a: $\sigma^2_{males} \neq \sigma^2_{females}$
   F = 3.805, p-value = 0.0037
   At α = 0.05, df_{num} = 20, df_{den} = 21, F_{cutoff} = 2.42 (two-tail test)
   Reject the null hypothesis. There is enough evidence to conclude that the variance in study time on weeknights differs between males and females.

Chapter 11 Extra Exercises - Solutions

Section 11.2 Exercises

1.   a) and c)

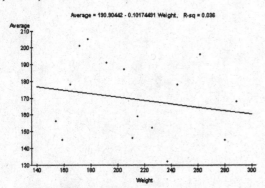

      b) $\hat{y} = 190.9 - 0.1017x$
      d) $190.9 - 0.1017(190) = 171.6$.
            The negative slope suggests that heavier bowlers tend to have lower averages. However, due to the scattered
            appearance of the data on the plot, the relationship does not appear to be very strong.
      e)   23.78

3.   a) and c)

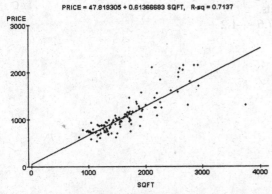

      b) $\hat{y} = 47.8193 + 0.6137x$
      d) $47.8193 + 0.6137(2200) = 1397.96*100 = \$139,796$
            The positive slope indicates that as the square footage increases, the selling price tends to increase. The graphs
            show a strong linear pattern, with a prominent outlier.
      e)   3750 square feet, sold for \$129,500

5.   a) and c)

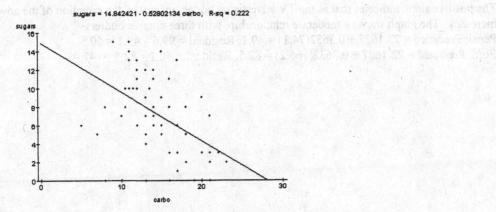

b) $\hat{y} = 14.8424 - 0.5280x$

d)   The negative slope suggests that as the number of grams of carbohydrates increases, the number of grams of sugar tends to decrease. The scatter plot suggests that the relationship is moderate.
As the number of carbohydrates increases by one gram, expect a decrease in the number of sugar grams by 0.53 gram.

e) $14.8428 - 0.5280(16) = 6.4$ grams of sugar

7.   a) and c)

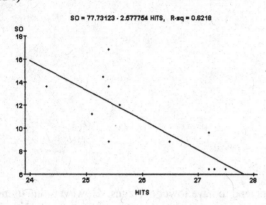

b) $\hat{y} = 77.7312 - 2.5778x$

d) $77.7312 - 2.5778(26) = 10.7$ strike outs
There is a negative slope, which indicates that as the number of hits increase, the number of strike outs tend to decrease. The scatter plot shows a moderate to strong relationship.

e)   Predicted = $77.7312 - 2.5778(27.2) = 7.6$ strike outs.
P. Derringer: $9.6 - 7.6 = 2$, F. Fitzsimmons: $6.4 - 7.6 = -1.2$

9.   a) and c)

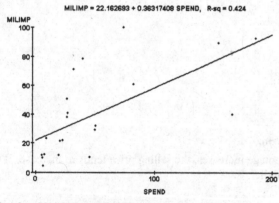

b) $\hat{y} = 22.1627 + 0.3632x$

d) $22.1627 + 0.3632(50) = 40.32$ million
The positive slope indicates that as the TV advertising budget increases, the retention of the advertisement increases. The graph shows a moderate relationship, with three notable outliers.

e)   Pepsi: Predicted = $22.1627 + 0.3632(74.1) = 49.1$, Residual = $99.6 - 49.1 = 50.5$
Ford: Predicted = $22.1627 + 0.3632(166.2) = 82.5$, Residual = $40.1 - 82.5 = -42.4$

11.   a) and c)

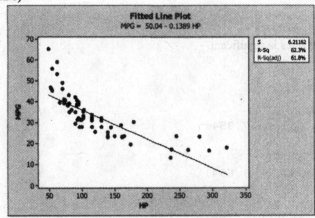

b) $\hat{y} = 50.04 - 0.139x$

d)   Expect a decrease of 0.14 miles per gallon in gas mileage, for each additional horse power of the engine. The negative slope suggests that as the horse power of the engine increases, the gas mileage will tend to decrease. The graph shows more of a curved relationship than a linear pattern.

e) $50.04 - 0.139(200) = 22.26$ miles per gallon

Section 11.3 Exercises

1.   a) $H_0: \beta_1 = 0$ $H_a: \beta_1 \neq 0$
        b) At $\alpha = 0.05$ with df = 13 for two-tails, $t_{cutoff} = +/- 2.16$.
        c) Slope coef = -0.1017, SE slope = 0.1459, t = -0.70
        d) The relationship between these two variables is not significant.

3.   a) $H_0: \beta_1 = 0$ $H_a: \beta_1 \neq 0$
        b) At $\alpha = 0.01$ with df = 115 for two-tails, $t_{cutoff} = +/- 2.58$.
        c) Slope coef = 0.61367, SE slope = 0.03625, t = 16.93
        d) The relationship between these two variables is significant.

5.   a) $H_0: \beta_1 = 0$ $H_a: \beta_1 \neq 0$
        b) At $\alpha = 0.02$ with df = 75 for two-tails, $t_{cutoff} = +/- 2.38$.
        c) Slope coef = 0.02576, SE slope = 0.00517, t = 4.98
        d) The relationship between these two variables is significant.

7.   a) $H_0: \beta_1 = 0$ $H_a: \beta_1 \neq 0$
        b) At $\alpha = 0.05$ with df = 17 for two-tails, $t_{cutoff} = +/- 2.11$.
        c) Slope coef = 0.8224, SE slope = 0.1739, t = 4.73
        d) The relationship between these two variables is significant.

9.   a) $H_0: \beta_1 = 0$ $H_a: \beta_1 \neq 0$
        b) At $\alpha = 0.01$ with df = 105 for two-tails, $t_{cutoff} = +/- 2.58$.
        c) Slope coef = 0.068904, SE slope = 0.004015, t = 17.16
        d) The relationship between these two variables is significant.

11.   a) $H_0: \beta_1 = 0$ $H_a: \beta_1 \neq 0$
        b) At $\alpha = 0.05$ with df = 79 for two-tails, $t_{cutoff} = +/- 1.99$.
        c) Slope coef = 0.238705, SE slope = 0.007082, t = 33.70
        d) The relationship between these two variables is significant.

13.
    a) H_0: $\beta_1 = 0$ H_a: $\beta_1 \neq 0$
    b)   At $\alpha = 0.05$ with df = 12 for two-tails, t_{cutoff} = +/- 2.179
    c)   Slope coef = 1.2673, SE slope = 0.4742, t = 2.67
    d)   The relationship between these two variables is significant.

Section 11.4 Exercises

1.
    a) 95.0% CI   b) 95.0% PI
       (30.232,   33.255)   (25.633,   37.854)

3.
    a) 99.0% CI   b) 99.0% PI
       (1.031,   1.670)   (-0.992,   3.694)

5.
    a) 95.0% CI   b) 95.0% PI
       (5.340,   6.018)   (4.163,   7.195)

7.
    a) 98.0% CI   b) 98.0% PI
       (3.6054,   3.8617)   (2.4608,   5.0063)

9.
    a) 99.0% CI   b) 99.0% PI
     (119.038, 121.481)   (110.622, 129.898)

11.
    a) 90.0% CI   b) 90.0% PI
       (46.82,   53.18)   (37.68,   62.32)

Section 11.5 Exercises

1.
    a) r = -0.1898
    b) There is a weak negative association between the two variables.
    c) $r^2 = 0.0360$

3.
    a) r = 0.8448
    b. There is a fairly strong positive association between the two variables.
    c. $r^2 = 0.7137$

5.
    a) r = 0.4986
    b) There is a moderate positive association between the two variables.
    c) $r^2 = 0.2486$

7.
    a) r = 0.7539
    b) There is a fairly strong positive association between the two variables.
    c) $r^2 = 0.5684$

9.
    a) r = 0.8586
    b) There is a fairly strong positive association between the two variables.
    c) $r^2 = 0.7372$

11.
    a) r = 0.9665
    b) There is a very strong positive association between the two variables.
    c) $r^2 = 0.9342$

13.
    a) r = 0.6124
    b) There is a moderate positive association between the two variables.
    c) $r^2 = 0.3750$

Section 11.6 Exercises

1. a)

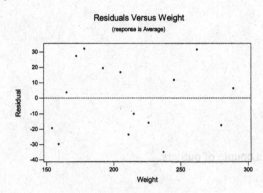

b) Yes. The points are scattered.
c) It is really hard to tell because the residuals seem to have a pattern.
d)

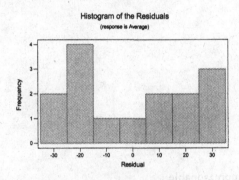

No. The graph shows a bimodal appearance.
e) No, since the residuals are not normal and r is very low.

3) a)

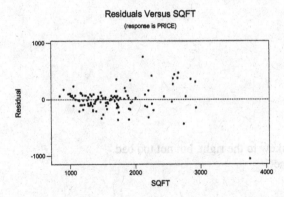

b) Yes. The points on the graph are scattered with the exception of an outlier.
c) No, the residuals seem to fan out.

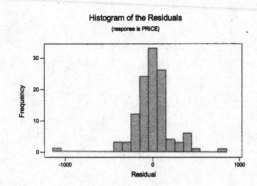

d)   Yes. The graph is bell shaped with the exception of a couple of outliers.
e)   No, the variance is a problem.

5.   a)

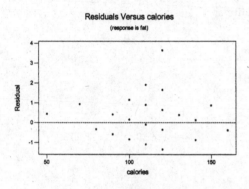

b)   Yes, the points on the graph are pretty scattered.
c)   The assumption of equal variances does not seem unreasonable.
d)

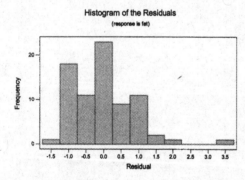

Yes, the curve is reasonably normal. There is a slight skew to the right, but not too bad.
e)   Yes, the graphs indicate that the assumptions are met.

7.   a)

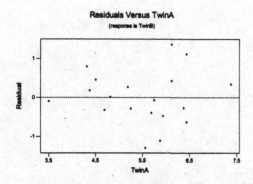

b)   Yes, the points on the graph are pretty scattered.
c)   No, the variance is clearly increasing as X increases.
d)

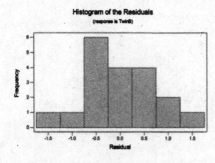

Yes, the graph is bell shaped.
e)   No, the equal variance assumption does not appear to hold.

9.   a)

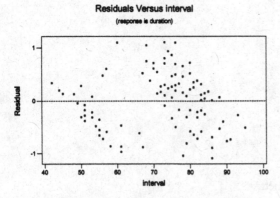

b)   Yes, the points on the graph are reasonably scattered.
c)   The equal variance assumption is not unreasonable.

d) Yes, the graph is fairly bell shaped.

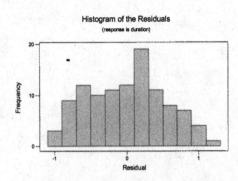

e) Yes, the graphs indicate that the assumptions are met.

11.   a)

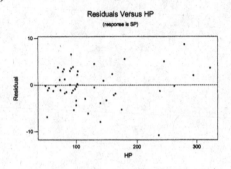

b) Yes, the points on the graph are reasonably scattered.
c) The equal variance assumption seems reasonable.
d) Yes, the graph is fairly bell shaped.

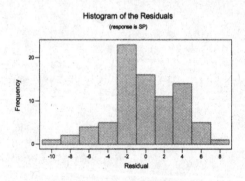

e. Yes, the graphs indicate that the assumptions are met.

13. a)

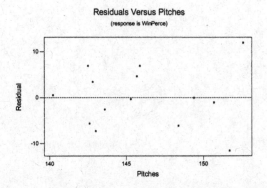

b) Yes, the points on the graph are reasonably scattered.
c) The equal variances assumption seems reasonable.
d) Yes, the graphs is bell shaped.

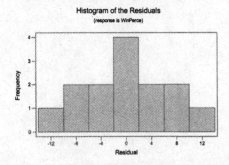

e) Yes, the graphs suggest that the assumptions are met.

Chapter 12 Extra Exercises - Solutions

Section 12.2 Exercises

1. a) Dependent: Number of bars of soap; Independent: Coupons Redeemed and Proportion on Sale

   b) $\hat{y} = 6.524 + 0.7008x_1 + 10.905x_2$

   c) For every extra coupon redeemed number of bars increases by 0.7008 and for every increase of 0.01 (1%) in proportion on sale the number increases by 0.10905.

   d. Expected sales $= 6.524 + (0.7008)(20) + 10.905(0.65)$
   $$= 6.524 + 14.016 + 7.09$$
   $$= 27.6 \approx 28 \text{ bars}$$

   e) This value seems reasonable because both of the values of the independent variables are within the range observed and the y value is within the range too. You are interpolating.

   f)

Observation	Predicted Y	Residuals
1	31.43	−0.429
2	25.65	−1.651
3	17.30	−0.302
4	12.55	1.448
5	13.47	−0.471
6	24.56	1.439
7	13.22	−3.221
8	31.63	1.369
9	18.66	−1.657
10	30.18	−2.183
11	34.72	−1.715
12	38.61	3.391
13	8.21	2.794
14	17.63	0.371
15	11.18	0.819

The models does a decent job of predicting. The largest error is just over 3 bars of soap and the smallest is less than 1 bar.

Section 12.3 Exercises

1. a)

Source	df	SS	MS	F
Regression	3	274.29	91.43	40.3
Residual	29	65.78	2.27	
Total		340.07		

   b) $H_0: \beta_1 = \beta_2 = \beta_3 = 0$
   H_A: At least one coefficient is not equal to 0.

   c) The critical value for the test is $F_{0.05, 3, 29} = 2.934$. Since 40.3 is outside this value we reject H_0 and conclude that the model is significant (at least one coefficient is not 0)

   d) The coefficient of determination is 80.7%.

3.   a)

Source	df	SS	MS	F
Regression	2	98.58	49.29	7.35
Residual	10	67.11	6.71	
Total	12	165.69		

$F = 7.34$, p value $= 0.0109$
Reject the null hypothesis. Conclude that at least one of the two coefficients is not equal to zero.

b)   The coefficient of determination is 59.5%
c)   The model is not bad, but it does not explain much more than half of the variation.
d) $t = -0.0008284 \div 0.7223 = -0.00115$, p value $= 0.991$

Do not reject the null hypothesis. Conclude that the first coefficient is equal to zero. Jordan's assists did not decrease the number of points he scored per game.

$t = 5.643 \div 1.889 = 2.987$, p value $= 0.0136$

Reject the null hypothesis. Conclude that the second coefficient is not equal to zero. Jordan's steals increased the number of points he scored per game.

e)   Based on the results I would suggest looking at other variables to see if there might be other factors that affected Michael Jordan's average points per game. The R^2 value is not high enough.

Chapter 13 Extra Exercises - Solutions

Section 13.3 Exercises

1. $[8(21.2) + 9(22.5) + 10(23.6) + 11(24.5)] \div 38 = 23.09$ (p. 698)

3. Grand mean = $(11.3 + 16.2 + 9.5 + 17.1) \div 4 = 13.525$

Sample Mean	20 $\times$ (Sample Mean – Grand Mean)^2
11.3	99.0125
16.2	143.1125
9.5	324.0125
17.1	255.6125
SSA	821.7500

5. $A = 5$; $B = 6$; $C = 90$; $D = 6$; $E = 1$

7. 107.0

9. 18.0

11. a)

SUMMARY

Groups	Count	Sum	Average	Variance
Column 1	5	80.3	16.06	6.598
Column 2	5	36.5	7.3	26.965
Column 3	5	12.8	2.56	3.373
Column 4	5	32.7	6.54	5.308
Column 5	5	47.3	9.46	14.528

ANOVA

Source of Variation	SS	df	MS	F	P-value	F crit
Between Groups	492.866	4	123.216	10.852	7.65E-05	2.866081
Within Groups	227.088	20	11.354			
Total	719.954	24				

b) H_0: $\mu_1 = \mu_2 = \mu_3 = \mu_4 = \mu_5$

c)

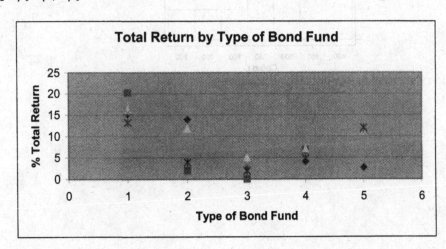

d)   Reject the null hypothesis. Conclude that at least one of the population means is different from the others. (pp. 702 – 707)

Section 13.4 Exercises

1.   a)   No, there are no obvious reasons to believe the errors are not independent.
     b)

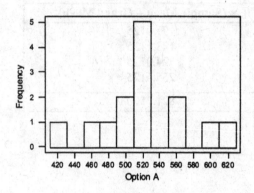

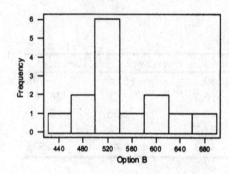

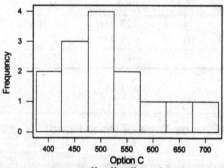

The plot for option C does not appear normally distributed.

c)

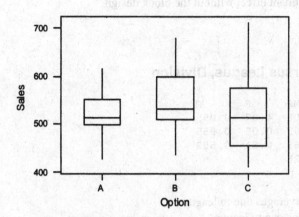

Option	Sample Variance
A	2555.96
B	4340.34
C	8389.21

The variances do not appear to be equal.

d) The assumption of independent errors appears to be the only valid assumption.

Section 13. 5 Exercises

1. a)

Group	Count	Sum	Average	Variance
Row 1	5	38.5	7.70	39.665
Row 1	5	39.8	7.96	64.528
Row 1	5	50.9	10.18	19.042
Row 1	5	44.6	8.92	23.737
Row 1	5	35.8	7.16	25.943
Column 1	5	80.3	16.06	6.598
Column 1	5	36.5	7.30	26.965
Column 1	5	12.8	2.56	3.373
Column 1	5	32.7	6.54	5.308
Column 1	5	47.3	9.46	14.528

ANOVA

Source of Variation	SS	df	MS	F	P-value	F crit
Between Groups	492.866	4	123.216	9.9171	0.0003	3.007
Between Blocks	28.294	4	7.073	0.5693	0.6887	3.007
Within Groups	198.794	16	12.425			
Total	719.954	24				

b) Relative Efficiency $= \dfrac{4(7.073) + 5(4)(123.216)}{24(12.425)} = 8.359$

c)   Although the variance due to the treatment is relatively small, the relative efficiency indicates that 8 times as many observations would be needed to detect the treatment effect without the block design.

Section 13.6 Exercises

1.   a) **Two-way ANOVA: Batting Average versus League, Division**

```
Source          DF          SS          MS      F      P
League           1   0.0005808   0.0005808   4.07   0.055
Division         2   0.0000131   0.0000065   0.05   0.955
Interaction      2   0.0001736   0.0000868   0.61   0.552
Error           24   0.0034220   0.0001426
Total           29   0.0041895
```

b) H_0: There is no difference in the population averages due to league.
     H_A: The is a difference in the population means due to league.
     Fail to reject H_0. League is not a significant factor in batting average.

H_0: There is no difference in the population means due to division..
H_A: There is a difference in the population means due to division.
Do not reject H_0, Division is not a significant factor in batting average.

H_0: There is no difference in the population means due to interaction between league and division.
H_A: There is a difference in the population means due to interaction between league and division.
Do not reject H_0, There is not a significant interaction between league and division.

c)

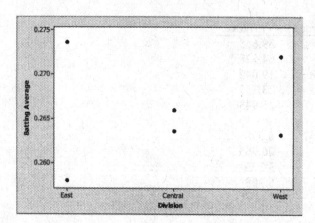

Chapter 14 Extra Exercises

Section 14.2 Exercises

1. a)

Week Day	M	T	W	TH	F
Messages	57	53	52	45	38
Expected	49	49	49	49	49

   b) The observed data seem to agree with the expected frequency, so it looks like he is correct.
   c) H_0: The messages are uniformly distributed over the weekdays
      H_A: The messages are not uniformly distributed over the weekdays.
   d) The chi-square statistic is 4.612. The critical value is 9.488. We cannot reject the null hypothesis.
   e) Based on the test, the claim is valid and the messages are uniformly distributed.

3. a)

Coupons Inserted	Number Observed	Expected $p = 30\%$	$(o - e)^2/e$
0	70	67.24	0.11666
1	150	144.08	0.25
2	110	123.48	1.496
3	55	52.92	0.09167
4	10	11.32	0.1286
5	5	0.96	22.05

   b) Chi-square = 24.13
   c) Since the critical value with 5 degrees of freedom is 11.070 we reject H_0. The data do not appear to be consistent with a binomial distribution with p = 0.30.

5. a) 38.7
   b)

Grade	K	1	2	3	4	5	6
Roster	33	39	38	42	40	38	41
Expected	38.7	38.7	38.7	38.7	38.7	38.7	38.7

   c) H_0: The distribution of pupils per grade is uniform.
      H_A: The distribution of pupils per grade is not uniform.
      The value of the test statistic is 1.19
   d) Since the critical value is 12.592 we cannot reject H_0. The principal can tell the PTA that the distribution of pupils per grade is uniform.

Section 14.3 Exercises

1. a) H_0: $p_1 = p_2 = p_3$
      H_A: at least one proportion is different
   b) 0.44, 0.35, 0.29 The proportions look like they might be different
   c) The value of the chi-square statistic is 2.140.
      Expected Values
      87.34   23.45   8.21
      61.66   16.55   5.79
   d) The p value is 0.343 so we cannot reject H_0. It appears that there is no difference in the proportions by racial group.

3. a)   0.32   0.42   0.52   0.70   0.85

b)

 0.54

c) Chi-square value is 40.333. Reject H_0. There is a difference in proportion for different education levels.

Section 14.4 Exercises

1. a) H_0: Year and type of computer owned are independent
   H_A: Year and type of computer owned are not independent

   b)
   ```
   Expected counts are printed below observed counts
   ```

	Freshman	Sophomore	Junior	Senior	Total
Laptop	111	115	117	157	500
	122.66	129.15	134.68	113.52	
Desktop	187	201	205	167	760
	186.44	196.31	204.71	172.54	
Both	67	78	82	65	292
	71.63	75.42	78.65	66.29	
Neither	145	143	156	83	527
	129.28	136.12	141.95	119.65	
Total	510	537	560	472	2079

   c) The test statistic is 37.357. There are 9 degrees of freedom so the critical value is 16.919. Since 35.357 is greater
   d) than 16.919 we reject H_0.
   d) We can conclude that year in college and type of computer owned are not independent.

3. a) H_0: Race and attitude toward marijuana legalization are independent
   H_A: Race and attitude toward marijuana legalization are not independent.

   b)
   ```
   Expected counts are printed below observed counts
   ```

	LEGAL	NOT LEGAL	Total
White	425	979	1404
	412.25	991.75	
Black	67	197	264
	77.52	186.48	
Other	33	87	120
	35.23	84.77	
Total	525	1263	1788

   c.) The value of the chi-square statistic is 2.779. The degrees of freedom are (3-1)*(2-1) = 2 so the critical value is 5.991. We cannot reject H_0 so we conclude that there is not enough evidence to say that race and attitude toward marijuana legalization are dependent.